Designing and Building
Electronic Filters

Designing and Building Electronic Filters

Delton T. Horn

TAB BOOKS
Blue Ridge Summit, PA

FIRST EDITION
FIRST PRINTING

© 1992 by **TAB Books**.
TAB Books is a division of McGraw-Hill, Inc.

Printed in the United States of America. All rights reserved. The publisher takes no responsibility for the use of any of the materials or methods described in this book, nor for the products thereof.

Library of Congress Cataloging-in-Publication Data

Horn, Delton T.
   Designing and building electronic filters / by Delton T. Horn.
     p. cm.
   ISBN 0-8306-3933-0 (p)
   1. Electric filters—Design and construction. I. Title.
TK7872.F5H67   1992b
621.381′5324—dc20   91-34957
                                              CIP

TAB Books offers software for sale. For information and a catalog, please contact TAB Software Department, Blue Ridge Summit, PA 17294-0850.

Acquisitions Editor: Roland S. Phelps
Book Editor: Laura J. Bader
Managing Editor: Sandra L. Johnson
Director of Production: Katherine G. Brown
Book Design: Jaclyn J. Boone
Cover Design: Graphics Plus, Hanover, PA.   EL2

Contents

Preface *vii*

❖ 1 **Passive low-pass filters** *1*

Frequency *1*
Basic filter types *2*
Passive and active filter circuits *5*
Practical versus ideal filters *6*
Capacitive reactance *8*
RC time constants *12*
The basic passive low-pass filter circuit *14*
Advantages and disadvantages of the basic low-pass filter *18*

❖ 2 **Other passive filters** *21*

The basic passive high-pass filter *21*
Filter symbols *25*
The basics of the band-pass filter *26*
The Q factor *31*
The passive band-pass filter *33*
The band-reject filter *41*
The basic passive band-reject filter circuit *46*
Comparison of filter types *49*

❖ 3 **Active low-pass filters** *55*

Basics of operational amplifiers *56*
The Butterworth low-pass filter *62*
The Chebyshev low-pass filter *65*
Infinite gain multiple feedback low-pass filter circuits *66*
VCVS low-pass filter circuits *73*
High-order filters *86*

- **4 Active high-pass filters** — **89**
  - Noise and active high-pass filters *92*
  - Infinite gain multiple feedback high-pass filter *93*
  - VCVS high-pass filter circuits *99*

- **5 Active band-pass filters** — **107**
  - More about Q *108*
  - Infinite gain multiple feedback band-pass filters *109*
  - The VCVS band-pass filter *120*
  - High-Q VCVS band-pass filter circuits *126*
  - High-order band-pass filters *132*

- **6 Active band-reject filters** — **133**
  - The passive twin-T band-reject filter *135*
  - The active twin-T band-reject filter circuit *138*
  - The bridged differentiator *142*
  - The active subtraction-type notch filter *149*
  - The VCVS active band-reject filter *154*

- **7 State variable and all-pass filters** — **161**
  - The basics of the state variable filter *161*
  - The unity gain state variable filter *163*
  - The four op amp state variable filter *168*
  - State variable band-reject filters *176*
  - The all-pass filter *185*

- **8 Voltage-controlled filters** — **201**
  - Basics of voltage control *205*
  - Regeneration and resonance *207*
  - Design of voltage-controlled filters *208*
  - Low-pass VCF *209*
  - Band-pass voltage-controlled filter *213*

- **9 Equalizers** — **219**
  - The band equalizer *219*
  - Parametric equalizers *222*

- **10 Digital filtering** — **223**
  - Digital filter circuits *223*
  - Advantages of software filters *227*
  - Digital signals *228*
  - A/D and D/A conversion *230*
  - Software filters *233*

- **Index** — **235**

Preface

TODAY ELECTRONIC CIRCUITS CAN BE AND ARE DESIGNED FOR almost any imaginable purpose. The possibilities seem virtually endless. However, in practice, most real electronic systems, no matter how complex, are ultimately made up of a number of simpler basic circuits. Many of these basic circuits can be used in a wide variety of ways. Most real-world electronic systems are made up of relatively simple circuits, such as amplifiers, oscillators, power supplies, and filters.

This book is concerned with filter circuits. This volume introduces you to filter theory and practice. After reading this book, you will be well on your way towards designing a suitable filter circuit for almost any electronic application. This book covers almost every imaginable type of filter circuit, ranging from simple passive filter networks to sophisticated digital filters.

A *filter* is a frequency sensitive circuit. It passes some frequencies, while blocking others. A *filter circuit* is defined according to which frequencies it passes and which it rejects. This book covers the four basic types of filter circuits (low-pass, high-pass, band-pass, and band-reject) and a number of specialized filter types.

Because of the nature of this kind of circuitry, quite a bit of math is required. I've made a great effort to make this necessary math as painless as possible. Specific examples are given for all the equations used in this book. If you can operate a simple "four-banger" calculator, you will have no problem with the vast majority of these equations.

I strongly encourage you to experiment with the filter circuits discussed here. At least one set of typical component values are given for all of the important circuits to get you started.

Whether you're an electronics novice or an old hand at electronics work, I think you'll agree that this book really does cover building electronic filters for every purpose.

❖ 1
Passive low-pass filters

THE TYPE OF ELECTRONIC CIRCUIT KNOWN AS A FILTER DERIVES ITS name from the mechanical filter and serves an analogous function. A simple, familiar mechanical filter is the paper cone found in a coffeemaker. This filter's purpose is to remove the grounds from the brewed coffee. It is made from a special type of paper that does not weaken or dissolve when the hot liquid is poured through it. A very porous type of paper is used for the filter. This means that there are many tiny, microscopic holes through the filter paper. The coffee can drip through these tiny holes, but the grounds are too large to fit through the holes. The freshly brewed coffee is passed through the filter to the cup, while the undesired grounds are held back in the filter.

Essentially, a mechanical filter "sorts" particles by size. Small particles can pass through the filter, while larger particles are blocked.

Frequency

In concept, an electronic filter works in a similar manner, except you are no longer dealing with particles but with electrical signals. Specifically, an electronic filter "sorts" or separates signals by their frequency. (For the rest of this book, when I use the word filter, an electronic filter is assumed unless otherwise specified.)

Electrical signals can be either ac (alternating current) or dc (direct current). A dc voltage has a specific, consistent value. An ac voltage, on the other hand, continuously oscillates from positive to negative and back again. (Sometimes an ac signal can "ride" on a dc signal, so it never passes through true zero.) These

Fig. 1-1 *The sine wave is the simplest of all ac waveforms.*

oscillations are in the form of a regular, repeating pattern. The simplest type of ac signal is the sine wave, illustrated in Fig. 1-1.

Each repetition of the pattern is called a cycle. The cycles are normally repeated at a regular rate, each cycle lasting a specific period of time. This repetition rate is the frequency of the signal. Frequency is measured as the number of complete cycles that occur in 1 second. The standard unit of measurement for frequency is the hertz (Hz). In some sources, frequency might be given in cps (cycles per second). This term is most likely to be found in older electronics texts. Don't let the different terminology confuse you, 1 Hz equals 1 cps.

Most ac signals used in electronic circuits have frequencies of several thousand hertz or more. For convenience, larger units are sometimes used. These larger frequency units use the standard metric prefixes. The two most commonly used are kilohertz (kHz) and megahertz (MHz). One kilohertz is equal to 1 000 hertz. One megahertz is equal to 1 000 000 hertz, or 1 000 kilohertz.

Of course, most signals won't have a tidy frequency value with a whole number of complete cycles each second. Fractional frequencies are common. For example, a signal with a frequency of 375.5 Hz has 375.5 cycles in 1 second's time. For our purposes, a dc signal can be considered as an ac signal with a frequency of 0 Hz.

For most electronic circuits, such as amplifiers, treat all signals equally regardless of their frequency. Any preferential treatment for certain frequencies is considered undesirable, and practical electronic circuits are usually designed to minimize such effects. A filter, however, is a special type of electronic circuit that is designed to be frequency selective in a specific and predictable way; that is, certain frequency signals are amplified more (or attenuated less) than other frequency signals.

Basic filter types

A filter circuit is designed for one (or sometimes more) specific cutoff frequency. This is the crossover point of the filter's action.

On one side of the cutoff frequency the signal is fully passed (receives maximum amplification or minimum attenuation), while on the other side of the cutoff frequency the signal is fully blocked (receives minimum amplification or maximum attenuation). A practical filter can't achieve perfect cutoff, but has a gradual crossover slope. I discuss this point shortly.

There are many different types of filters, depending on the specific intended application. In practical terms, however, just four basic types are in common use. These four basic filter types are the low-pass filter, the high-pass filter, the band-pass filter, and the band-reject filter. Each of these names describes the action of the filter.

A low-pass filter passes low-frequency signals, but blocks or rejects high-frequency signals. Figure 1-2 shows a frequency response graph for an ideal low-pass filter. Any signal with a frequency higher than the filter's cutoff frequency is blocked, while any signal with a frequency lower than the filter's cutoff frequency is passed.

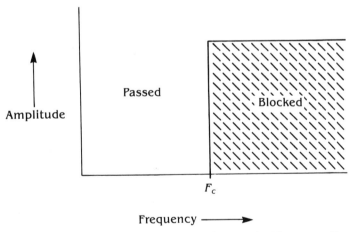

Fig. 1-2 A frequency response graph for an ideal low-pass filter.

A high-pass filter, as you might suspect, is just the opposite of a low-pass filter. This is illustrated in the frequency response graph for an ideal high-pass filter, shown in Fig. 1-3. In a high-pass filter, any signal with a frequency higher than the filter's cutoff frequency is passed, while any signal with a frequency lower than the filter's cutoff frequency is blocked.

A frequency response graph for an ideal band-pass filter is shown in Fig. 1-4. In this type of filter, only signals within a spe-

4   Passive low-pass filters

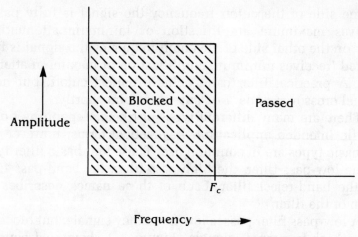

Fig. 1-3 *A frequency response graph for an ideal high-pass filter.*

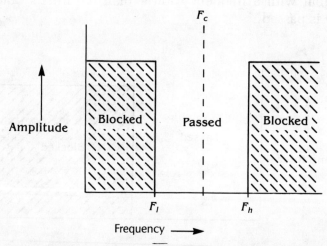

Fig. 1-4 *A frequency response graph for an ideal band-pass filter.*

cific range or band of frequencies are passed. Any signal with a frequency outside the passband (either too high or too low) is blocked.

As its name suggests, a band-reject filter is the opposite of a band-pass filter. All frequency signals are passed except those falling within a specific band or range. A frequency response graph for an ideal band-reject filter is illustrated in Fig. 1-5. Because of the appearance of this frequency response graph, a band-reject filter is sometimes called a notch filter. A "notch" is

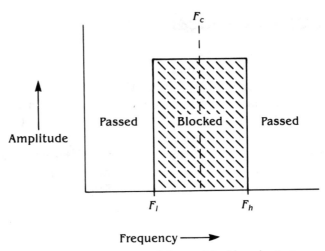

Fig. 1-5 A frequency response graph for an ideal band-reject or notch filter.

made in the otherwise (nominal) flat frequency response of the circuit.

I discuss high-pass filters, band-pass filters, and band-reject filters in more detail in chapter 2, but for the remainder of this chapter I focus on the low-pass filter.

Passive and active filter circuits

A filter circuit can be either passive or active. A passive filter is made up entirely of passive components, specifically resistors, capacitors, and sometimes inductors. It requires no power supply. In effect, the passive filter circuitry "steals" its operating power from the signal being passed through it. This means that the entire signal is attenuated to some degree. Blocked frequencies are attenuated much more than passed frequencies. A passive filter circuit is not capable of amplification.

An active filter circuit, on the other hand, includes an amplifier stage. In addition to attenuating blocked frequencies, an active filter also boosts or amplifies passed frequencies. Obviously active filter circuits must be built around some sort of active electronic components, such as transistors, tubes, or integrated circuits. All active filter circuits require a power supply of some sort.

We begin our examination of active filter circuits in chapter 3. For the time being, concentrate on the simpler passive filter circuit.

Practical versus ideal filters

In the frequency response graphs of Figs. 1-2 through 1-5, we are assuming ideal filter performance. For a low-pass filter, any signal frequency that is slightly higher than the cutoff frequency is completely blocked, while any signal frequency that is slightly below the cutoff frequency is completely passed. Such ideal filtering is impossible to achieve in practice. Any practical filter circuit is going to have an intermediate range or slope. Frequency components close to the cutoff frequency are partially passed and partially blocked. This is shown in Fig. 1-6, which is a frequency response graph for a typical practical low-pass filter.

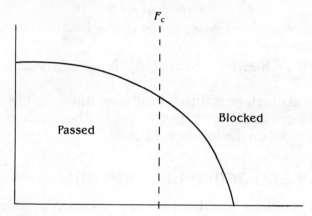

Fig. 1-6 *A frequency response graph for a typical practical low-pass (or other) filter.*

Frequencies near the cutoff frequency (including all frequencies slightly higher than and slightly lower than this nominal value) are partially attenuated. They are neither completely passed nor completely blocked. In a low-pass filter, frequencies below the cutoff frequency are not attenuated as much as higher frequencies. The higher the signal frequency, the greater the attenuation. Frequency components significantly lower than the cutoff frequency are fully passed, and frequency components significantly higher than the cutoff frequency are fully blocked (or greatly attenuated).

The filter's slope is a measure of how rapidly the frequency components near the cutoff frequency go from the passed condition to the blocked state. Obviously, the better the filter is the steeper its frequency response slope will be.

The ideal low-pass filter of Fig. 1-2 has a crossover slope of zero. At the cutoff frequency, the signal switches instantly from fully passed to fully blocked. Of course, such an infinite slope is impossible to achieve with any practical circuitry.

In practical filter circuits, the crossover slope is defined as x dB/octave. A decibel (dB) is a comparative value indicating the difference between the measured quantity and a static reference value. In this case, the reference value is the full, unattenuated (passed) amplitude of the original input signal. The measured quantity is the amplitude of the output signal after the relevant frequency component has traveled through the filter circuit.

It is important to remember that the decibel is not a linear value; that is, an 8-dB signal is not twice as strong as a 4-dB signal. Instead, the decibel indicates a logarithmic scale, as shown in the graph of Fig. 1-7. The values being compared are related exponentially rather than linearly. This corresponds to the way the human ear hears changes in amplitude.

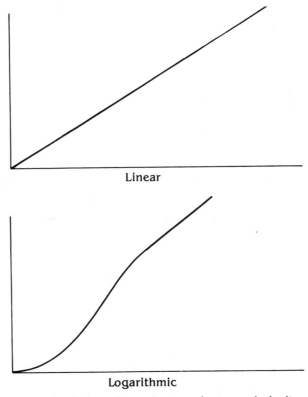

Fig. 1-7 *The decibel is a logarithmic scale, instead of a linear scale.*

In case you aren't familiar with the term octave, it represents a doubling of frequency. For example, 200 Hz is one octave higher than 100 Hz, and 400 Hz is one octave higher than 200 Hz (and two octaves higher than 100 Hz). Notice that higher octaves are larger than lower octaves.

The decibel per octave measurement of a filter's crossover slope tells how much the signal amplitude is reduced for each doubling of frequency. A simple passive filter circuit might have a gradual slope of just 3 dB/octave. Most active filter circuits feature crossover slopes of at least 6 dB/octave. Better filter circuits typically have slope ratings of 12 dB/octave, 24 dB/octave, or even higher.

The slope decibel value doesn't necessarily have to be an exact multiple of three, but in practical circuits it usually will be. This is because of certain core similarities in almost all standard filter circuits. Basically, an active filter circuit incorporates a passive filter network and multiplies its effect. A standard passive filter network's slope is 3 dB/octave, so an active filter circuit built around such a network will almost always have a slope that is some multiple of 3 dB/octave.

I discuss crossover slopes and their significance in more detail once we get to our study of active filter circuits.

Capacitive reactance

Many advanced filter circuits are quite sophisticated, but simple filtering can be accomplished with very little circuitry. In fact, the simplest possible filter consists of just a single component—a capacitor. All capacitors function as high-pass filters. They block low-frequency signals while allowing high-frequency signals to pass through them. The chief function of a capacitor is to block dc voltages while passing ac voltages. A dc voltage is the same as an ac signal with a frequency of 0 Hz.

A capacitor by itself is not really a functional filter, but you need to understand the principles involved in capacitive filtering action before you can truly understand actual filter networks. So how does the capacitor "know" which frequency components to pass and which to block? The answer is an electrical parameter known as capacitive reactance, a form of ac resistance.

Ordinary dc resistance is frequency independent. It doesn't matter if the applied signal is dc (0 Hz) or ac with a frequency of 30 Hz or 50 000 Hz—the dc resistance has the same effect on all signals. An ac resistance, such as capacitive reactance, is frequency sensitive. Some frequency components experience more effective resistance than others. In the case of capacitive reactance, the resistance decreases as the signal frequency increases.

Reactance is expressed in ohms (like ordinary dc resistance), but the ohmic reactance value of any specific component varies with the applied signal frequency. Capacitive reactance is determined by the frequency of the applied signal and the capacitance of the component being used. The formula for capacitive reactance is

$$X_c = \frac{1}{2\pi FC}$$

where
  F = applied signal frequency, in hertz;
  C = the capacitance, in farads; and
  π = 3.14

This particular formula has 2π, which is equal to 6.28. This means the formula for capacitive reactance can be rewritten as

$$X_c = \frac{1}{6.28 FC}$$

First, assume you are using a 100-μF (0.000 1-F) capacitor. If you apply a 100-Hz signal across this capacitor, the capacitive reactance will be equal to

$$X_c = \frac{1}{6.28 \times 0.000\,1 \times 100}$$
$$= \frac{1}{0.062\,8}$$
$$= 15.923\,566\,9\ \Omega$$

If you increase the signal frequency to 5 000 Hz, the capacitive reactance changes to

$$X_c = \frac{1}{6.28 \times 0.000\,1 \times 5\,000}$$
$$= \frac{1}{3.14}$$
$$= 0.318\,471\,3\ \Omega$$

Conversely, if you reduce the signal frequency to 35 Hz, the capacitive reactance goes up to

$$X_c = \frac{1}{6.28 \times 0.000\,1 \times 35}$$

$$= \frac{1}{0.021\,98}$$

$$= 45.495\,905 \ \Omega$$

As you can see, increasing the signal frequency decreases the capacitive reactance, while decreasing the signal frequency increases the capacitive reactance.

What if a dc signal is applied across a capacitor? Any dc signal has an effective frequency of 0 Hz, so the capacitive reactance in this case becomes

$$X_c = \frac{1}{6.28 \times 0.000\,1 \times 0}$$

$$= \frac{1}{0}$$

$$= \infty$$

Division by zero always results in infinity. For a dc signal, the capacitive reactance is (theoretically) infinitely large. This is always true, regardless of the capacitance of the component used.

With any practical capacitor, true infinity can never be achieved. There will always be some leakage (dc) resistance through the capacitor. The amount of leakage resistance will depend on various factors in the construction of the specific capacitor. The leakage resistance will normally be quite large, so very little of the dc signal can get through the capacitor. For most practical purposes, it is reasonable to assume that the resistance to the dc signal is infinite. The error will usually be negligible and of little practical significance.

Try a few more examples of capacitive reactance. This time assume you are using a smaller capacitor. With a capacitance value of 0.025 μF (0.000 000 025 F), if the applied signal has a frequency of 100 Hz, the capacitive reactance works out to

$$X_c = \frac{1}{6.28 \times 0.000\,000\,025 \times 100}$$

$$= \frac{1}{0.000\,015\,7}$$

$$= 63\,694.267 \; \Omega$$

Increasing the applied signal frequency to 5 000 Hz drops the capacitive reactance to

$$X_c = \frac{1}{6.28 \times 0.000\,000\,025 \times 5\,000}$$

$$= \frac{1}{0.000\,785}$$

$$= 1\,273.885\,3 \; \Omega$$

Decreasing the applied signal frequency to 35 Hz changes the capacitive reactance to

$$X_c = \frac{1}{6.28 \times 0.000\,000\,025 \times 35}$$

$$= \frac{1}{0.000\,005\,4}$$

$$= 185\,185.18 \; \Omega$$

The capacitive reactance is inversely proportional to the capacitance and to the applied signal frequency. Increasing either the capacitance or the frequency decreases the capacitive reactance and vice versa. Some additional examples of capacitive reactance are given in Table 1-1.

Table 1-1 Examples of capacitive reactance.

Frequency (Hz)	C = 0.01 μF	C = 0.02 μF	C = 0.05 μF	C = 0.1 μF
100	159 236	79 618	31 847	15 924
200	79 618	39 841	15 924	7 962
300	53 079	26 539	10 616	5 308
500	31 847	15 924	6 369	3 185
1 000	15 924	7 962	3 185	1 592
1 500	10 616	5 308	2 123	1 062
2 500	6 369	3 185	1 274	637
5 000	3 185	1 592	637	319
7 000	2 275	1 137	455	228
10 000	1 592	796	318	159
12 500	1 274	637	255	127
47 000	339	169	68	34

RC time constants

By itself, a capacitor is a kind of filter, but in practical circuits the capacitor is almost always combined with a resistor. Resistor-capacitor combinations are one of the most important types of networks in electronic circuits. Such a combination of a resistor and a capacitor is known as an RC network or RC circuit.

If a resistor and a capacitor are connected in series across a voltage, as illustrated in Fig. 1-8, the capacitor will be charged through the resistor at a specific rate determined by both the capacitance and the resistance. The capacitance determines how many electrons the negative plate of the capacitor can hold when fully charged, and the resistance slows down the flow of electrons.

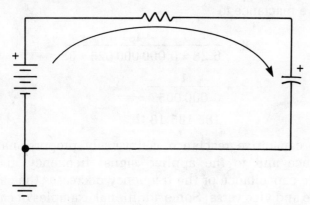

Fig. 1-8 *If a resistor and a capacitor are connected in series across a voltage, the capacitor will be charged through the resistor at a specific rate determined by both the capacitance and the resistance.*

The time it takes for the capacitor to be charged (from zero) to 63% of its full potential charge level is called the time constant of the RC combination. Similarly, if the voltage is removed from the circuit, as in Fig. 1-9, the capacitor will be discharged through the resistor. In this case, the time constant is defined as the time it takes for the capacitor to drop (from fully charged) to 37% of its fully charged value. The charging time constant and the discharging time constant are always equal for any given combination of a specific capacitance and resistance.

The use of the 63% and 37% points in defining the time constant of an RC network might seem rather strange and arbitrary. To an extent it is an arbitrary choice, but one that was made for a

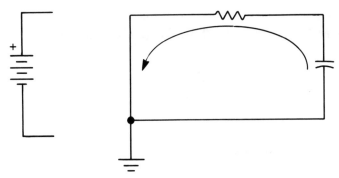

Fig. 1-9 *If the voltage is removed from the circuit, the capacitor will be discharged through the resistor.*

very good reason. By defining the time constant in this seemingly odd manner, the required mathematical equations are made as simple as possible. The time constant of any specific RC circuit can be found simply by multiplying the resistance and the capacitance:

$$T = RC$$

where
    T = time constant, in seconds;
    R = resistance, in megohms (1 MΩ = 1 000 000 Ω), and
    C = capacitance, in microfarads (μF).

If you use R in ohms, then C must be in farads. Always be careful not to confuse the units of measurement, or the equation won't come out correctly.

For example, if the resistor has a value of 100 000 Ω (0.1 MΩ) and the capacitor is 10 μF, then the time constant will be

$$T = 0.1 \times 10$$
$$= 1 \text{ s}$$

Note that the same time constant could be achieved with other RC combinations. For instance, a 1-MΩ resistor and a 1-μF capacitor would also give a time constant of 1 second. So would a 10-Ω resistor and a 100-μF capacitor.

As another example, let's say we have a 470 000-Ω (0.47-MΩ) resistor in series with a 2.2-μF capacitor. In this case, the time constant works out to

$$T = 0.47 \times 2.2$$
$$= 1.034 \text{ s}$$

Let's try one more example. This time we will use a 22 000-Ω (0.022-MΩ) resistor and a 0.3-μF capacitor, giving us a time constant of

$$T = 0.022 \times 0.3$$
$$= 0.006\,6 \text{ s}$$
$$= 6.6 \text{ ms}$$

As you can see, the time constant of an RC network is always quite short, usually well under 1 second.

The basic passive low-pass filter circuit

The basic low-pass filter circuit is essentially a variation on the simple RC network discussed in the preceding section. The resistor is still in series with the capacitor, but the capacitor is arranged so that it is in parallel with both the input and the output. This simple circuit is shown in Fig. 1-10.

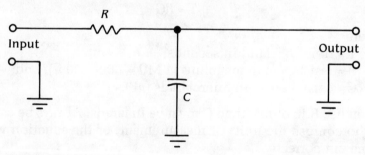

Fig. 1-10 *The basic low-pass filter circuit is essentially a variation on the simple RC network shown in Fig. 1-8.*

The entire input signal (all frequency components) passes through the resistor. A capacitor passes high frequencies, but blocks low frequencies. If the input signal contains a high-frequency component, most of its amplitude will be shunted to ground through the capacitor and effectively lost from the output signal. A low-frequency component in the input signal, however, can't pass through the capacitor, so it has no place to go except to the output. As a result, the output signal contains the low-frequency components from the original input signal, but not the high-frequency components. In other words, this simple circuit acts like a low-pass filter.

The slope of such a passive filter circuit is very shallow, just 3 dB/octave, but it is a functional filter and sufficient for many practical purposes. Because of the shallow crossover slope, there is usually plenty of leeway in the component values. There is rarely much point in using expensive high-precision components. Standard component values that are reasonably close to the calculated values will usually work just as well.

The formula for finding the cutoff frequency for this type of simple passive low-pass filter is

$$F_c = \frac{159\,000}{RC}$$

where
F_c = nominal cutoff frequency, in hertz;
R – resistance, in ohms; and
C = capacitance, in microfarads.

Once again, we will work through a few practical examples to give you a more concrete feel for how this type of filter circuit works. First, assume you are using a 47-kΩ (47 000-Ω) resistor and a 0.022-μF capacitor. The cutoff frequency of this filter would work out to

$$F_c = \frac{159\,000}{47\,000 \times 0.022}$$

$$= \frac{159\,000}{1\,034}$$

$$= 154 \text{ Hz}$$

As a second example, use an 18-kΩ (18 000-Ω) resistor and a 0.33-μF capacitor. This time the cutoff frequency is

$$F_c = \frac{159\,000}{18\,000 \times 0.33}$$

$$= \frac{159\,000}{5\,940}$$

$$= 27 \text{ Hz}$$

In the third example, use a 2.7-kΩ (2 700-Ω) resistor with a 0.004 7-μF capacitor, giving a nominal cutoff frequency of

$$F_c = \frac{159\,000}{2\,700 \times 0.004\,7}$$

$$= \frac{159\,000}{12.69}$$

$$= 12\,530 \text{ Hz}$$

Increasing the resistance (R) or the capacitance (C) (or both) decreases the nominal cutoff frequency. Some additional examples are given in Table 1-2.

In most practical electronics work, you will probably already know the desired cutoff frequency and will need to determine the required component values. This is easy to do. First, select a likely capacitance value for C, then rearrange the basic equation to solve for R instead of F:

$$R = \frac{159\,000}{FC}$$

As an example, suppose you need a passive low-pass filter with a cutoff frequency of 1 700 Hz (1.7 kHz). Try using a 0.01-μF capacitor. In this case, you would need a resistor with a value of

$$R = \frac{159\,000}{1\,700 \times 0.01}$$

$$= \frac{159\,000}{17}$$

$$= 9\,353 \text{ }\Omega$$

Obviously this is not a standard resistor value, but the gradual slope of this type of passive filter gives you plenty of leeway in rounding off the component values. You can almost always use the nearest standard resistance value in either direction. That is, you can either round down to 9.1 kΩ (9 100 Ω) or up to 10 kΩ (10 000 Ω). Because 10-kΩ resistors are much more common, that is the choice I would make.

If you are concerned that the actual cutoff frequency might be too far off from the desired value, you can plug the rounded component values back into the original equation and double-check your work. For the example, you find that the nominal cutoff frequency has been changed to

$$F = \frac{159\,000}{10\,000 \times 0.01}$$

$$= \frac{159\,000}{100}$$

$$= 1\,590 \text{ Hz}$$

Table 1-2 Typical RC combinations for passive low-pass filter circuits.

R (kΩ)	C (μF)	F (Hz)
1	0.01	15 900
2.2	0.01	7 227
3.3	0.01	4 818
4.7	0.01	3 383
6.8	0.01	2 338
10	0.01	1 590
22	0.01	723
47	0.01	338
100	0.01	159
1	0.022	7 227
2.2	0.022	3 285
3.3	0.022	2 190
4.7	0.022	1 538
6.8	0.022	1 063
10	0.022	723
22	0.022	329
47	0.022	154
100	0.022	72
1	0.033	4 818
2.2	0.033	2 190
3.3	0.033	1 460
4.7	0.033	1 025
6.8	0.033	709
10	0.033	482
22	0.033	219
47	0.033	103
100	0.033	71
1	0.047	3 383
2.2	0.047	1 538
3.3	0.047	1 025
4.7	0.047	720
6.8	0.047	497
10	0.047	383
22	0.047	154
47	0.047	72
100	0.047	50
1	0.1	1 590
2.2	0.1	723
3.3	0.1	482
4.7	0.1	338
6.8	0.1	234
10	0.1	159
22	0.1	72
47	0.1	34
100	0.1	16

In a practical passive filter circuit, this isn't likely to give any noticeable difference compared to a true 1 700-Hz cutoff frequency. Also remember that the component tolerances will almost certainly throw off the actual result at least a little from the calculated value. Such errors are generally negligible in passive filter circuits because of the very gradual crossover slope. If by some chance you do come up with an unreasonable resistance value (such as 0.87 Ω or 29 548 000 Ω), you can simply select a new capacitor value and try the resistance equation again.

In some applications it may be desirable to replace the fixed resistance (R) with a variable resistance, such as with a potentiometer. It is usually a good idea to use a fixed resistor in series with the potentiometer (or other variable-resistance device) to limit the minimum resistance seen by the circuit.

Advantages and disadvantages of the basic low-pass filter

The basic low-pass filter circuit has several valuable advantages. It is very simple, both to design and to construct. It is also quite inexpensive, requiring just two simple components—one resistor and one capacitor. The price for both components is likely to come in at well under one dollar.

Because it is a purely passive circuit, it doesn't require any kind of power supply. This circuit takes its operating power from the signal it is filtering. This brings us to one of the major disadvantages of the simple passive low-pass filter. All of the signal is attenuated to some extent, even frequency components well below the nominal cutoff frequency. The entire signal must pass through the resistor causing an all-frequency voltage drop in addition to the filtering action of the capacitor. If the original input signal is relatively weak, the output level might be too low to satisfactorily drive the next stage in the system. If this "across the board" attenuation is a problem in your application, a passive filter is not sufficient for your needs. You will need to use an active filter that includes an amplifier to boost the passed signal. I discuss active filter circuits beginning in chapter 3.

Another disadvantage of the basic passive low-pass filter circuit is the very gradual crossover slope. The filtering action is just fair, at best, as indicated in the frequency response graph of Fig. 1-11. In some cases, you can steepen the crossover slope by

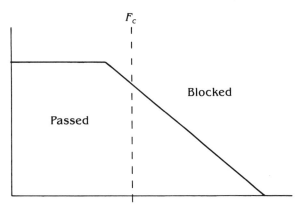

Fig. 1-11 *A major disadvantage of the basic passive low-pass filter circuit is its very gradual crossover slope.*

cascading two or more identical passive filter networks in series, as shown in Fig. 1-12. If the signal is passed through two identical filters, each frequency component will be subjected to twice as much filtering. For example, suppose the first filter reduces a given frequency component's level by one unit. (It doesn't matter what the unit is.) The second filter stage is identical to the first and it also reduces the frequency component by one unit, so in the final output signal the frequency component will be attenuated by two units.

Fig. 1-12 *In some cases, you can steepen the crossover slope by cascading two or more identical passive filter networks in series.*

A single passive low-pass filter stage has a crossover slope of 3 dB/octave. Two such circuits in series have a combined effective crossover slope of 6 dB/octave. This steepened slope comes at a price, however. Remember, any passive filter reduces the entire signal's amplitude. If the signal passes through two filter stages, the overall attenuation effect will be doubled. If a steeper crossover slope is required without excessive attenuation of the passed signal, you will have to use an active filter circuit instead of a simple passive filter circuit.

❖ 2
Other passive filters

THE LOW-PASS FILTER IS PROBABLY THE MOST COMMON TYPE, BUT many practical applications call for other filtering operations. Chapter 1 looked at the basic low-pass filter. In this chapter I examine simple passive filter circuits for the other three basic filter functions—high-pass filters, band-pass filters, and band-reject filters.

The basic passive high-pass filter

Figure 2-1 shows the basic low-pass filter circuit made up of just two simple components—a capacitor and a resistor. High-frequency components are shunted to ground through the capacitor.

What would happen if you reversed the positions of these two components, as shown in Fig. 2-2? In this case, any low-frequency components in the input signal will be blocked (or severely attenuated) by the capacitor, so only the high-frequency components will be passed through to the output. In other words, you now have a high-pass filter circuit. The frequency response graph for this type of filter is shown in Fig. 2-3.

A high-pass filter can be considered a "mirror image" of a low-pass filter. The operation works in exactly the opposite way. Assuming that the cutoff frequency is the same for both the low-pass filter and the high-pass filter, any frequency component that is blocked by the low-pass filter will be passed by the high-pass filter and any frequency component that is passed by the low-pass filter will be blocked by the high-pass filter.

If a passive low-pass filter circuit and a passive high-pass filter circuit use the same capacitor and resistor values, both filter

22   Other passive filters

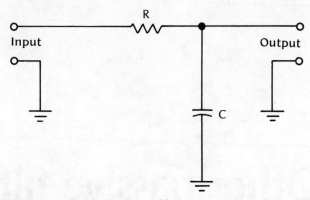

Fig. 2-1 The basic passive low-pass filter circuit is made up of a simple resistor and a capacitor.

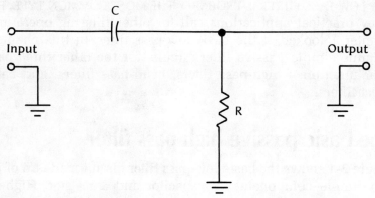

Fig. 2-2 For a passive high-pass filter circuit, the positions of the resistor and the capacitor are reversed.

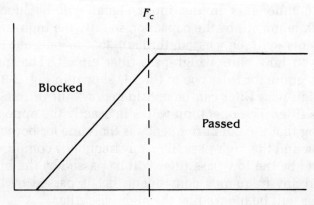

Fig. 2-3 The frequency response graph for a typical passive high-pass filter.

circuits will have the same cutoff frequency. In other words, the formula for determining the cutoff frequency of a passive high-pass filter is exactly the same as the cutoff frequency formula for the passive low-pass filter circuit:

$$F = \frac{159\,000}{RC}$$

where
    F = filter's nominal cutoff frequency, in hertz;
    R = resistance, in ohms; and
    C = capacitance, in microfarads.

In working with this formula (or any other equation in electronics work), always be careful to use the correct units for all values. If you express R in kilohms and C in picofarads, you will get an incorrect result.

Like the basic low-pass filter circuit discussed in chapter 1, the basic high-pass filter circuit has a very shallow slope of about 3 dB/octave. For this reason, exact component values and tolerances usually aren't critical. If you need a cutoff frequency of 2 500 Hz (2.5 kHz), but the result of the equation is 2 789 Hz (2.789 kHz), the difference will probably be negligible. It might not even be noticeable in practical operation.

Try a couple of practical examples of this formula. First, assume you have a simple passive high-pass filter circuit made up of a 22-kΩ (22 000-Ω) resistor and a 0.003 3-μF capacitor. What is the nominal cutoff frequency of this filter? All you have to do is plug these component values into the equation and solve for F:

$$F = \frac{159\,000}{RC}$$

$$= \frac{159\,000}{22\,000 \times 0.003\,3}$$

$$= \frac{159\,000}{72.6}$$

$$= 2\,190 \text{ Hz}$$

You can round this off and say you have a high-pass filter with a cutoff frequency of 2.2 kHz (2 200 Hz) or even 2 kHz (2 000 Hz).

As a second example, try using a 390-kΩ (390 000-Ω) resistor and a 0.022-μF capacitor. In this case, the nominal cutoff frequency of your passive high-pass filter works out to

$$F = \frac{159\,000}{390\,000 \times 0.022}$$

$$= \frac{159\,000}{8\,580}$$

$$= 18.5 \text{ Hz}$$

Notice that increasing either the resistance (R) or the capacitance (C) decreases the nominal cutoff frequency.

Of course, in a practical design situation, you will probably know the desired nominal cutoff frequency and will need to determine the proper component values to achieve that particular cutoff frequency (or come close to it). To design a passive high-pass filter circuit for a specific cutoff frequency, simply select a likely capacitor value (C), then algebraically rearrange the formula to solve for the unknown resistance (R) instead of for the known frequency (F):

$$R = \frac{159\,000}{FC}$$

For example, suppose you need a passive high-pass filter circuit with a cutoff frequency of about 1.7 kHz (1 700 Hz). You might try using a 0.01-μF capacitor. In this case, the necessary resistance value (R) works out to

$$R = \frac{159\,000}{1\,700 \times 0.01}$$

$$= \frac{159\,000}{17}$$

$$= 9\,353 \text{ }\Omega$$

This is not a standard resistance value, but it is close to 9.1 kΩ (9 100 Ω), which is a standard value for 5% tolerance resistors. In some applications a common 10-kΩ (10 000-Ω) resistor might be close enough.

But assume you cannot find a 9.1-kΩ resistor and 10 kΩ isn't close enough. You could use a trimpot to fine tune the resistance to the desired value, or you could substitute a new capacitance value (C) and rework the equation to solve for the new value of R. Try a 0.005-μF capacitor this time. Now you need a resistor with a value of

$$R = \frac{159\,000}{1\,700 \times 0.005}$$

$$= \frac{159\,000}{8.5}$$

$$= 18\,706\ \Omega$$

In this case you can use an 18-kΩ (18 000-Ω) resistor or possibly a more common 22-kΩ (22 000-Ω) resistor.

Of course, if necessary, unusual resistance values can be obtained by combining multiple resistors in series or in parallel. Generally speaking, this will probably not be necessary when working with simple filter circuits like those discussed in this chapter and in chapter 1. This is because the very shallow cutoff slope of such filter circuits permits very wide tolerances in the actual component values used.

Filter symbols

In block diagrams, a filter section is commonly indicated by a small equilateral triangle with a small, simplified frequency response graph in it to identify the type of filter action performed by the section. The filter triangle is normally positioned on edge, with the input fed into the wide base and the output taken off the angle tip on the opposite end. Symbols for low-pass and high-pass filters are shown in Fig. 2-4. Notice that the same symbol is used for either a passive or an active filter. You will need these symbols when I discuss the concepts involved in the remaining two basic filter types—the band-pass filter and the band-reject filter.

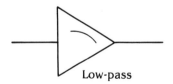

Fig. 2-4 The symbols used to indicate a low-pass filter and a high-pass filter.

The basics of the band-pass filter

Suppose you place a low-pass filter network in series with a high-pass filter network, as illustrated in Fig. 2-5. (For convenience in this part of the discussion, I assume all frequency components are either fully blocked or fully passed.) Any frequency component blocked by either (or both) of the filter sections will be deleted from the final output signal. Only those frequency components that are passed by both filter sections will reach the output.

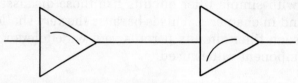

Fig. 2-5 *If we place a low-pass filter network in series with a high-pass filter network, the result will be a simple band-pass filter.*

I call the cutoff frequency of the low-pass filter section A and the cutoff frequency of the high-pass filter section is identified as B. There are three possible relationships between these two cutoff frequencies:

- A less than B,
- A equal to B, and
- A greater than B.

As you shall see, only one of these three combinations results in a practical filter circuit. Remember I am assuming that we are dealing with ideal filters here. All frequency components are either fully passed or fully blocked. No partial attenuation is considered here.

Figure 2-6 shows the combined frequency graph for when A is less than B (the cutoff frequency of the low-pass filter section is lower than the cutoff frequency of the high-pass filter section). The low-pass filter section passes everything below its cutoff frequency and blocks everything above this point. On the other hand, the high-pass filter section passes everything above its cutoff frequency and blocks any frequency component below this point. Because the low-pass cutoff frequency is lower than the high-pass cutoff frequency, the net result is that everything is

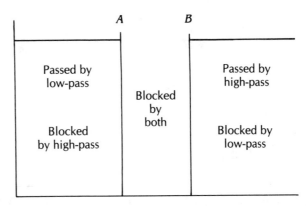

Fig. 2-6 *The combined frequency response graph for when A is less than B in the cascaded circuit of Fig. 2-5.*

blocked. None of the input signal is passed through to the output. This is clearly not a functional filter circuit. It blocks everything and the entire input signal might as well not exist.

You have a similar situation when A (low-pass cutoff frequency) and B (high-pass cutoff frequency) are equal, as illustrated in Fig. 2-7. The low-pass filter blocks everything below this point and the high-pass filter passes everything above this point, so the entire input signal is blocked and the output is completely dead. You still don't have a practical filter circuit.

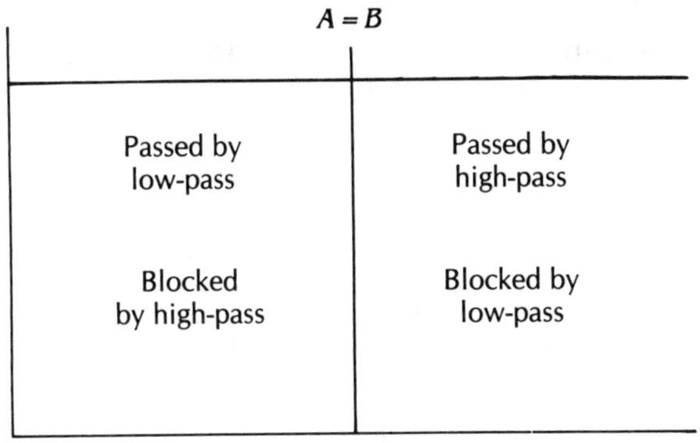

Fig. 2-7 *The combined frequency graph for when A equals B in the cascaded circuit of Fig. 2-5.*

The third possible combination, however, is functional. This time A (the low-pass cutoff frequency) is higher than B (the high-pass cutoff frequency). A typical frequency response graph for this combination is shown in Fig. 2-8. Here you have a band of

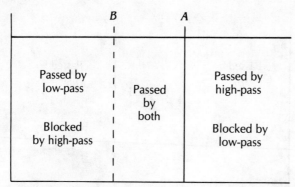

Fig. 2-8 *The combined frequency response graph for when A is greater than B in the cascaded circuit of Fig. 2-5.*

frequencies higher than B but lower than A that are passed by both of the series filter sections. Any frequency component higher than A is blocked by the low-pass filter and any frequency component lower than B is blocked by the high-pass filter, but all frequency components between B and A are passed.

As you can see from the frequency response graph, only those frequency components within a specific band are passed by this type of filter. Any frequency component outside the specified band is rejected or blocked. It should be obvious from looking at this frequency response graph why this type of filter is called a band-pass filter. The range of passed frequencies is called, not surprisingly, the passband.

Unlike the simple low-pass and high-pass filters discussed thus far, a band-pass filter has two cutoff frequencies to be considered. They are usually identified as F_l (lower cutoff frequency) and F_h (upper cutoff frequency). The lowest end of the passband is F_l and F_h marks the uppermost end of the passband. Notice that F_l is the same as what we have been calling B, and F_h is identical to what we have been calling A. The distance between F_l and F_h is called the bandwidth, because this is a measurement of how wide the passband is.

Usually when describing a band-pass filter, the actual cutoff frequencies (F_l and F_h) are not directly specified. Instead, the filter's bandwidth and center frequency are identified. The F_l and F_h points can easily be derived from these specifications. The center frequency (usually identified as F_c) is simply the midpoint of the passband (halfway between F_l and F_h).

In most practical applications, the bandwidth and center frequency will be of more direct interest and importance than the F_l and F_h cutoff frequencies. It is very easy to convert back and forth between these two sets of specifications. For instance, if you know the F_l and F_h cutoff frequencies, you can easily find the bandwidth (BW) simply by taking the difference between these two frequencies:

$$BW = F_h - F_l$$

The center frequency (F_c) can be found by taking the average between the lower and upper cutoff frequencies. The easiest way to do this is to add the two cutoff frequencies together and divide the sum by two:

$$F_c = \frac{F_h + F_l}{2}$$

Try this out with a practical example. Suppose you have a band-pass filter circuit with a lower cutoff frequency (F_l) of 1 000 Hz and an upper cutoff frequency (F_h) of 2 000 Hz. What are the bandwidth and center frequency for this particular filter?

First, take the difference of the two cutoff frequencies to find the bandwidth of the filter:

$$BW = F_h - F_l$$
$$= 2\,000 - 1\,000$$
$$= 1\,000 \text{ Hz}$$

Next, take the average of the filter's center frequency:

$$F_c = \frac{(F_h + F_l)}{2}$$
$$= \frac{2\,000 + 1\,000}{2}$$
$$= \frac{3\,000}{2}$$
$$= 1\,500 \text{ Hz}$$

The example filter has a bandwidth of 1 000 Hz centered around 1 500 Hz.

You can just as easily work in the opposite direction and determine the lower and upper cutoff frequencies (F_l and F_h)

from the bandwidth and center frequency specifications. For example, suppose you have a band-pass filter circuit with a bandwidth of 800 Hz and a center frequency of 650 Hz. What are the actual cutoff frequencies of this particular filter?

By definition, half of the bandwidth is below the center frequency and the other half is above the center frequency. To find the lower cutoff frequency (F_l), you simply subtract one-half the bandwidth (BW) from the center frequency (F_c):

$$F_l = F_c - \frac{BW}{2}$$

Similarly, adding one-half the bandwidth (BW) to the center frequency (F_c) gives the upper cutoff frequency of the filter:

$$F_h = F_c + \frac{BW}{2}$$

For the example filter, the lower cutoff frequency works out to

$$F_l = 650 - \frac{800}{2}$$
$$= 650 - 400$$
$$= 250 \text{ Hz}$$

The upper cutoff frequency of this particular filter works out to

$$F_h = 650 + \frac{800}{2}$$
$$= 650 + 400$$
$$= 1\,050 \text{ Hz}$$

In other words, the passband of the filter in the example runs from 250 to 1 050 Hz.

Thus far, I have assumed ideal filtering action. All frequency components are either fully blocked or fully passed. However, such ideal filtering is impossible to achieve with any practical circuitry. Simple passive filter circuits won't even come close. Such ideal filters are useful for understanding the basic concepts involved. But any practical band-pass filter will have a finite cutoff slope, just like any practical low-pass or high-pass filter circuit. The frequency response graph for a typical band-pass filter is shown in Fig. 2-9. Notice that the slope is identical on both ends of the passband; that is, the slope of the

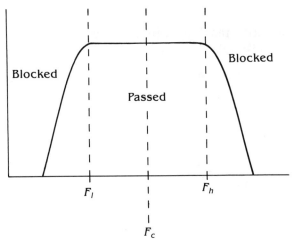

Fig. 2-9 *The frequency response graph for a typical band-pass filter.*

lower cutoff frequency (F_l) is exactly as steep as the slope of the upper cutoff frequency (F_h).

In a sense, all standard filters can be considered variants of the band-pass filter. A low-pass filter has a passband that extends below zero, while a high-pass filter has a passband that reaches above the highest useable frequency. A band-reject filter (discussed later in this chapter) is just an inverted form of the basic band-pass filter.

The Q factor

An important specification when dealing with filter circuits (especially band-pass filters and band-reject filters) is the Q or quality factor, which can be given for any capacitor, inductor, or tuned circuit. Basically it is a figure of merit. A high Q factor indicates a low-loss, high-efficiency circuit or component. A lower Q rating means the circuit or component in question has greater loss and lower efficiency. The higher the Q factor, the lower the loss.

On the component level, Q is directly related to the component's reactance and dc resistance. For an inductor in a series circuit, Q is equal to:

$$Q = \frac{X_l}{R}$$

where
- X_l = inductive reactance, in ohms; and
- R = dc resistance, in ohms.

Similarly for a capacitor in a series circuit, Q is equal to:

$$Q = \frac{-X_c}{R}$$

where
- X_c = capacitive reactance, in ohms; and
- R = dc resistance, in ohms.

The minus sign in this equation indicates the difference in phase between an inductive reactance and a capacitive reactance. Notice that in either an inductor or a capacitor (in a series circuit), increasing the reactance or decreasing the dc resistance causes an increase in the Q factor.

For a tuned circuit, such as a filter, Q is defined by the resonant (or center) frequency and the bandwidth according to the simple formula

$$Q = \frac{F_c}{BW}$$

Both the center frequency (F_c) and the bandwidth (BW) are given in hertz.

To better understand how the Q factor works, try a few specific examples. Suppose you have a band-pass filter circuit with a center frequency of 1 200 Hz and a bandwidth of 500 Hz. The Q of this filter is equal to

$$Q = \frac{1\,200}{500}$$

$$= 2.4$$

If you keep the same center frequency (1 200 Hz) but increase the bandwidth to 900 Hz, the Q changes to

$$Q = \frac{1\,200}{900}$$

$$= 1.33$$

On the other hand, if you leave the bandwidth at 500 Hz but increase the center frequency to 2 600 Hz, the Q changes to

$$Q = \frac{2\,600}{500}$$
$$= 5.2$$

Increasing the center frequency or decreasing the bandwidth increases the Q of the filter. Conversely, Q can be decreased by decreasing the center frequency or increasing the bandwidth.

You should be aware that even though Q is called the "quality factor," a high Q value doesn't always indicate the best possible filter circuit for a specific application. In many practical applications, a fairly wide passband may be desirable or even essential. This means that in such cases, a low Q value is "better" than a high Q value. In other applications, you will want the narrowest possible passband, calling for the highest possible Q factor. It all depends on how the filter circuit is to be used.

The passive band-pass filter

As I have already discussed, a band-pass filter can be made up of a low-pass filter and a separate high-pass filter in series. This will work, but it is not a very elegant solution. In many practical applications, this approach would be rather awkward.

A more efficient and elegant solution is to use a dedicated band-pass filter circuit. Such a circuit can be created simply by adding an inductor (or coil) in parallel with the capacitor in a passive low-pass filter circuit. The basic passive band-pass filter circuit is shown in Fig. 2-10.

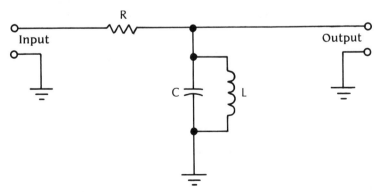

Fig. 2-10 *The basic passive band-pass filter circuit.*

The bandwidth of this filter is equal to what would be the low-pass cutoff frequency if the inductor were omitted from the circuit. In other words, you are using the same formula as before, but this time it gives you the filter's bandwidth:

$$BW = \frac{159\,000}{RC}$$

where
  BW = bandwidth, in hertz;
  R = value of the resistor, in ohms; and
  C = value of the capacitor, in microfarads.

Notice that the inductance (L) has no direct effect on the bandwidth of the filter. The center frequency, on the other hand, is defined by the capacitor (C) and inductor (L) values, but the resistance (R) is ignored. The center frequency of the filter is set by the resonant frequency of the parallel LC combination.

Resonance is an extremely important concept in electronics work, especially where tuned (frequency sensitive) circuits are involved. Resonance is directly related to reactance or ac resistance. There are two types of reactance: capacitive reactance and inductive reactance. These two types of reactance have opposite phase.

Capacitive reactance decreases as the signal frequency is increased. Inductive reactance, on the other hand, increases as the signal frequency is increased. For any LC combination there will be one (and only one) specific frequency at which the capacitive reactance is exactly equal to the inductive reactance. This is the resonant frequency of the LC combination.

The impedance of a circuit is the total ac resistance, made up of both types of reactance (capacitive and inductive) and the dc resistance. The formula for impedance is:

$$Z = \sqrt{R^2 + X_t^2}$$

where
  Z = impedance, in ohms;
  R = dc resistance, in ohms; and
  X_t = total effective reactance (both capacitive and inductive), in ohms.

Because of the phase difference between capacitive reactance and inductive reactance, these two values cannot simply be added together. In a series LC circuit, the total effective reactance (X_t) is equal to the difference between the inductive reactance and the capacitive reactance:

$$X_t = X_l - X_c$$

Notice that at resonance, the inductive reactance (X_l) is exactly equal to the capacitive reactance (X_c). These two factors cancel each other out, leaving a total effective reactance (X_t) of zero. This means that at resonance, a series circuit has an impedance of

$$Z = \sqrt{R^2 + 0^2}$$

$$= \sqrt{R^2 + 0}$$

$$= \sqrt{R^2}$$

$$= R$$

At the resonant frequency, a series resonant circuit has an impedance equal to its dc resistance. This is the minimum possible impedance for that particular circuit.

In the band-pass filter circuit, you have a parallel LC combination, so you need a somewhat different formula. In a parallel resonant circuit, the total effective reactance (X_t) is equal to

$$X_t = \frac{(X_l)(X_c)}{(X_l - X_c)}$$

At resonance, the inductive reactance (X_l) is equal to the capacitive reactance (X_c), so the denominator of this formula becomes zero:

$$X_t = \frac{(X_l)(X_c)}{0}$$

$$= \infty \text{ (infinity)}$$

Dividing any value by zero always results in infinity, so at resonance, a parallel resonant circuit has (theoretically) infinite reactance, and therefore (theoretically) infinite impedance. Obviously, this is the maximum possible impedance value for this particular circuit. In a practical circuit, true infinite impedance is not possible, but the impedance of a parallel LC circuit at resonance is extremely high, and much higher than at any other frequency.

The resonant frequency of any LC combination is defined by the capacitive reactance and the inductive reactance. These factors, in turn, are determined by the capacitance and the inductance of the components being used in the circuit.

Any combination of an inductance and a capacitance will have a resonant frequency. Any given LC combination will be resonant at only a single, unique frequency. The formula for finding the resonant frequency for a specific inductance-capacitance (LC) combination (either in a series circuit or in a parallel circuit) is:

$$F = \frac{1}{2\pi\sqrt{LC}}$$

where
 F = resonant frequency, in hertz;
 L = inductance, in henries; and
 C = capacitance, in farads.

The symbol π (pi) is a universal mathematical constant equal to approximately 3.14. Therefore, 2π is equal to about 6.28 and the formula can be rewritten as

$$F = \frac{1}{6.28\sqrt{LC}}$$

This resonant frequency is the center frequency of our passive band-pass filter circuit. Without worrying too much about the exact steps of the mathematical derivation, this formula can be rearranged to a somewhat more convenient form for our purposes:

$$F_c = \frac{159}{\sqrt{LC}}$$

where
 F_c = center (resonant) frequency, in hertz;
 L = inductance, in henries; and
 C = capacitance, in microfarads.

In most practical circuit design, these units will be the most convenient to work with, especially since the bandwidth formula also uses microfarads instead of farads for C.

Work your way through a few practical examples to become more familiar with these formulas. As the first example, assume you have a passive band-pass filter circuit with the following component values:

- R = 2.2 kΩ (2 200 Ω)
- C = 0.05 μF
- L = 50 mH (0.05 H)

In this particular filter circuit, you have a bandwidth of

$$BW = \frac{159\,000}{RC}$$

$$= \frac{159\,000}{2\,200 \times 0.05}$$

$$= \frac{159\,000}{110}$$

$$= 1\,445 \text{ Hz}$$

The bandwidth of this filter is just a little less than 1.5-kHz wide. Next, you can calculate the center frequency:

$$F_c = \frac{159}{\sqrt{LC}}$$

$$= \frac{159}{\sqrt{0.05 \times 0.05}}$$

$$= \frac{159}{\sqrt{0.002\,5}}$$

$$= \frac{159}{0.05}$$

$$= 3\,180 \text{ Hz}$$

The example passive band-pass filter circuit has a center frequency of 3 180 Hz and the bandwidth is 1 445 Hz. This means that all frequencies within 722.5 Hz (in either direction) will fall into the passband of this filter. In other words, this combination of components (ignoring the shallow slope of the passive filter circuitry) will filter out all frequency components except those between 2 457.5 and 3 902.5 Hz.

The Q of this filter is equal to

$$Q = \frac{F_c}{BW}$$

$$= \frac{3\,180}{1\,445}$$

$$= 2.2$$

As a second example, look at a passive band-pass filter circuit with the following component values:

- R = 10 kΩ (10 000 Ω)
- C = 0.22 μF
- L = 100 mH (0.1 H)

In this circuit, the bandwidth works out to

$$BW = \frac{159\,000}{10\,000 \times 0.22}$$

$$= \frac{159\,000}{2\,200}$$

$$= 72.27 \text{ H}$$

This filter's center frequency works out to

$$F_c = \frac{159}{\sqrt{0.1 \times 0.22}}$$

$$= \frac{159}{\sqrt{0.022}}$$

$$= \frac{159}{0.148}$$

$$= 1\,072 \text{ Hz}$$

The Q of our second example band-pass filter circuit is

$$Q = \frac{1\,072}{72}$$

$$= 14.89$$

Notice that this filter has a much narrower passband and a much higher Q factor than the filter of our previous example. Ignoring the effects of the shallow cutoff slope of the passive filter circuitry, this band-pass filter passes only those frequency components that fall between 1 036 and 1 108 Hz.

Of course, in most practical applications, you will already know what center frequency and bandwidth you want from your band-pass filter circuit, and you will need to select suitable component values that will work out to the desired parameters. This can be done by algebraically rearranging the bandwidth and center frequency equations.

The easiest approach is to arbitrarily select a likely capacitor value (C) and use the rearranged bandwidth equation to solve for the unknown resistance (R):

$$R = \frac{159\,000}{(BW)(C)}$$

The same value of C can then be used to determine the appropriate inductor value (L) in the rearranged center frequency equation:

$$L = \left(\frac{1}{C}\right)\left(\frac{159}{F}\right)^2$$
$$= \left(\frac{1}{C}\right)\left(\frac{159}{F}\right)\left(\frac{159}{F}\right)$$

As an example of the use of these modified equations, assume you need a band-pass filter circuit with a passband ranging from 1 200 to 2 700 Hz. The bandwidth of this filter is equal to

$$BW = F_h - F_l$$
$$= 2\,700 - 1\,200$$
$$= 1\,500 \text{ Hz}$$

The center frequency is the midpoint of the passband or the average of the two extremes. In this case, the center frequency works out to

$$F_c = \frac{F_h + F_l}{2}$$
$$= \frac{2\,700 + 1\,200}{2}$$
$$= \frac{3\,900}{2}$$
$$= 1\,950 \text{ }\Omega$$

In designing this band-pass filter circuit, begin by working out the low-pass (RC) filter section to give the desired bandwidth. Try it with a 0.022-μF capacitor:

$$R = \frac{159\,000}{(BW)(C)}$$

$$= \frac{159\,000}{1\,500 \times 0.022}$$

$$= \frac{159\,000}{33}$$

$$= 4\,818.181\,8 \ \Omega$$

This calculated resistance value is very close to 4 700 kΩ (4.7 Ω), which happens to be a very common standard resistor value.

The next step in the design procedure is to calculate the necessary inductance to give a center frequency reasonably close to 1 950 Hz:

$$L = \left(\frac{1}{C}\right)\left(\frac{159}{F}\right)^2$$

$$= \left(\frac{1}{0.022}\right)\left(\frac{159}{1\,950}\right)^2$$

$$= 45.454\,5 \times 0.081\,5^2$$

$$= 45.454\,5 \times 0.006\,65$$

$$= 0.302\,2 \ \text{H}$$

A 300-mH (0.30-H) coil should do just fine.

The parts list for your passive band-pass filter circuit with a center frequency of 1 950 Hz and a bandwidth of 1 500 Hz is as follows:

- R = 4.7 kΩ (4 700 Ω)
- C = 0.022 μF
- L = 300 mH (0.3 H)

For a second example, design another passive band-pass filter circuit, but one that will pass all frequency components from 700 to 3 600 Hz. In this case, the filter's bandwidth works out to

$$BW = 3\,600 - 700$$
$$= 2\,900 \ \text{Hz}$$

The center frequency for this filter works out to

$$F_c = \frac{3\,600 + 700}{2}$$

$$= \frac{4\,300}{2}$$

$$= 2\,150 \text{ Hz}$$

The Q of this filter will be

$$Q = \frac{F_c}{\text{BW}}$$

$$= \frac{2\,150}{2\,900}$$

$$= 0.74$$

In designing this second sample band-pass filter circuit, I arbitrarily selected a capacitor value of 0.1 μF. Solving for R first, you get

$$R = \frac{159\,000}{2\,900 \times 0.1}$$

$$= \frac{159\,000}{290}$$

$$= 548.28 \; \Omega$$

The nearest standard resistor value in this case is 560 Ω.

Next, calculate the appropriate value for the coil:

$$L = \left(\frac{1}{0.1}\right)\left(\frac{159}{2\,150}\right)^2$$

$$= 10 \times 0.74^2$$

$$= 10 \times 0.005\,47$$

$$= 0.054\,7 \text{ H}$$

A 50-mH (0.05-H) coil should probably be close enough considering the shallow cutoff slope of a simple passive filter circuit like this.

The band-reject filter

Earlier in this chapter we learned that placing a low-pass filter in series with a high-pass filter produces a band-pass filter. This is shown again in Fig. 2-11. What do you suppose would happen if you combined a low-pass filter and a high-pass filter in parallel rather than in series, as illustrated in Fig. 2-12?

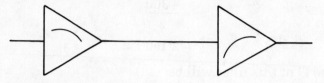

Fig. 2-11 *Placing a low-pass filter in series with a high-pass filter produces a band-pass filter.*

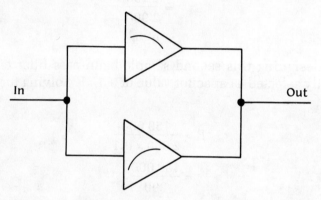

Fig. 2-12 *Combining a low-pass filter and a high-pass filter in parallel produces a band-reject filter.*

Once again, we call the cutoff frequency of the low-pass filter section A and the cutoff frequency of the high-pass filter section B. The three possible relationships between these two cutoff frequencies are:

- A greater than B,
- A equal to B, and
- A less than B.

Once more, as in the series combination, only one of these three possible combinations results in a practical filter circuit.

Because the two filter sections are in parallel, a given frequency component must be blocked by both filter sections or it will appear at the output. Any part of the input signal that passes through either filter section will appear in the output signal.

Once more, for convenience in the discussion, I assume that ideal filter circuits are being used. This permits us to ignore complications due to the cutoff slopes.

Figure 2-13 shows the combined frequency response graph for the case of A greater than B (the cutoff frequency of the low-pass filter section is higher than the cutoff frequency of the high-pass filter section). The low-pass filter section passes everything below its cutoff frequency (A). At the same time, the high-pass filter section passes everything above its cutoff frequency (B). Because A is higher than B, any frequency component blocked by the low-pass filter section will be passed by the high-pass filter section and vice versa. There is even some overlap where certain frequency components are passed by both filter sections. The parallel combination of the low-pass and high-pass filters will result in no filtering at all. The two filter sections cancel each other out.

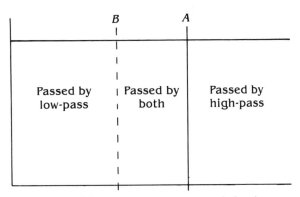

Fig. 2-13 *The combined frequency response graph for the case of A greater than B in the parallel filter circuit of Fig. 2-12.*

Figure 2-14 shows the effect when A equals B. This is similar to the A greater than B condition, except there is no overlapping band of frequency components passed by both filter sections. However, any individual frequency component that is not passed by the low-pass filter section will be passed by the high-pass filter section and vice versa. The absence of the overlap band makes absolutely no difference in the output signal, and the output signal will be the same as the input signal. Clearly, neither of these situations gives a practical filter circuit.

Now consider the third possibility where A is less than B. A typical frequency response graph for this situation is shown in

44 Other passive filters

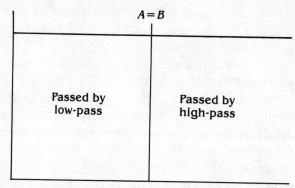

Fig. 2-14 *The combined frequency response graph for the case of A equals B in the parallel filter circuit of Fig. 2-12.*

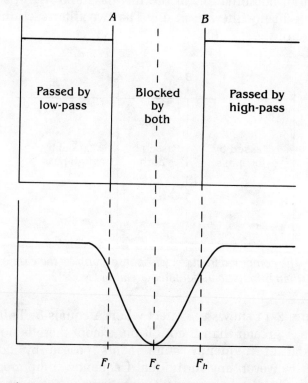

Fig. 2-15 *The combined frequency response graph for the case of A less than B in the parallel filter circuit of Fig. 2-12.*

Fig. 2-15. Because the low-pass cutoff frequency (A) is lower than the high-pass cutoff frequency (B) there is an overlapping blocked band. Any frequency components falling into this overlap band will be blocked by both of the individual filter sections and will therefore be filtered out of the final, combined output signal.

Notice that all of the input frequency components are passed by this parallel circuit, except for those frequency components within the overlapping blocked band. The net result is a mirror image of the band-pass filter discussed earlier. In this case, however, the identified band is not passed, it is blocked or rejected, and this type of filter circuit is called a band-reject filter.

As you can see from the frequency response graph of Fig. 2-15, the circuit's frequency response has a hole or notch. For this reason, the band-reject filter is often called the notch filter.

Like the band-pass filter, the operation of a band-reject filter can be defined by the lower cutoff frequency (B or F_l) and the upper cutoff frequency (A or F_h). It is more common, however, to describe the reject band or notch by its center frequency and bandwidth. These terms are exactly the same as with the band-pass filter. The bandwidth is the distance between the upper cutoff frequency (F_h) and the lower cutoff frequency (F_l):

$$BW = F_h - F_l$$

Similarly, the center frequency is the exact midpoint of the reject band or notch. It can be found by taking the average of the upper cutoff frequency (F_h) and the lower cutoff frequency (F_l):

$$F_c = \frac{F_h + F_l}{2}$$

These equations work in exactly the same way as for the band-pass filter discussed earlier in this chapter. Only the filtering action is reversed. What is passed by the band-pass filter is blocked by the band-reject filter and vice versa.

Let's run through one quick example here. Suppose you have a band-reject filter with a lower cutoff frequency (F_l) of 570 Hz and an upper cutoff frequency (F_h) of 1 340 Hz. The bandwidth of this particular filter is equal to

$$\begin{aligned}BW &= F_h - F_l \\ &= 1\,340 - 570 \\ &= 770 \text{ Hz}\end{aligned}$$

This filter's center frequency (F_c) works out to

$$F_c = \frac{F_h + F_l}{2}$$

$$= \frac{1\,340 + 570}{2}$$

$$= \frac{1\,910}{2}$$

$$= 955 \text{ Hz}$$

Band-reject filters use the Q factor just like band-pass filters, and it works in exactly the same way. The Q of a filter is the ratio of its center frequency to its bandwidth:

$$Q = \frac{F_c}{BW}$$

For the example band-reject filter circuit, the Q is equal to

$$Q = \frac{955}{770}$$

$$= 1.24$$

Of course, band-reject filter circuits with higher or lower Q factors are possible.

The basic passive band-reject filter circuit

Back at the beginning of this chapter, you learned that a high-pass filter is functionally the mirror image of a low-pass filter and that you can make a passive high-pass filter circuit by taking a passive low-pass circuit and reversing the positions of the components. Now you know that a band-reject filter is functionally the mirror image of a band-pass filter. So can you create a passive band-reject filter circuit simply by reversing the positions of a passive band-pass filter circuit? You sure can.

The basic passive band-pass filter circuit is shown in Fig. 2-16. This is basically just a passive low-pass filter circuit with an added inductor or coil in parallel across the capacitor.

Figure 2-17 shows the basic passive band-reject filter circuit. Compare this circuit with the one shown in Fig. 2-16. In this case you have a passive high-pass filter circuit (a passive low-pass filter circuit with the component positions reversed), and once again, an inductor or coil has been added to the circuit in parallel across the capacitor.

All of the equations for this passive band-reject filter network are exactly the same as for the passive band-pass filter cir-

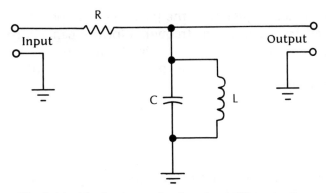

Fig. 2-16 *The basic passive band-pass filter circuit.*

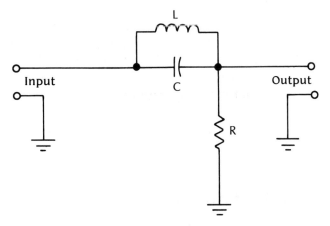

Fig. 2-17 *The basic passive band-reject filter circuit.*

cuit discussed earlier in this chapter. The only difference is that those frequency components that are blocked by the band-pass filter will be passed by the comparable band-reject filter and vice versa.

The bandwidth for this passive band-reject filter circuit is found using the same formula used to find the cutoff frequency of the high-pass filter, ignoring the coil altogether:

$$BW = \frac{159\,000}{RC}$$

where
 BW = bandwidth of the reject range, in hertz;
 R = value of the resistor, in ohms; and
 C = value of the capacitor, in microfarads.

The center frequency of the passive band-reject filter circuit is defined by the resonant frequency of the capacitor-inductor combination used in the circuit. This formula is the same as the center frequency used for the passive band-pass filter circuit, described earlier:

$$F_c = \frac{159}{\sqrt{LC}}$$

where
- F_c = filter's center frequency (the midpoint of the reject band), in hertz;
- L = inductance of the coil, in henries; and
- C = value of the capacitor, in microfarads.

Because these are the very same formulas used in the discussion of the basic band-pass filter circuit earlier in this chapter, I won't go into great detail here. I'll just go through one quick example. Assume you are working with a passive band-reject filter circuit with the following component values:

- $R = 4.7$ kΩ (4 700 Ω)
- $L = 150$ mH (0.15 H)
- $C = 0.033$ μF

The bandwidth of this particular filter is

$$\begin{aligned}
BW &= \frac{159\,000}{RC} \\
&= \frac{159\,000}{4\,700 \times 0.033} \\
&= \frac{159\,000}{155.1} \\
&= 1\,025.145 \text{ Hz}
\end{aligned}$$

For simplicity round this off to 1 000 Hz. Component tolerances and the passive filter circuitry's shallow cutoff slope will totally mask any functional difference due to this rounding off.

The next step is to find the center frequency of the example band-reject filter circuit:

$$F_c = \frac{159}{\sqrt{LC}}$$

$$= \frac{159}{\sqrt{(0.15 \times 0.033)}}$$

$$= \frac{159}{\sqrt{0.004\ 95}}$$

$$= \frac{159}{0.070\ 356\ 2}$$

$$= 2\ 260\ \text{Hz}$$

This particular filter circuit passes all frequency components except for those falling between 1 760 (F_l) and 2 760 Hz (F_h).

Finally, you can determine the Q of the sample passive band-reject filter circuit:

$$Q = \frac{F_c}{\text{BW}}$$

$$= \frac{2\ 260}{1\ 000}$$

$$= 2.26$$

If you understand how to work with the basic passive band-pass filter circuit, you should have no problems working with the basic passive band-reject filter circuit.

Comparison of filter types

Each of the four basic types of filters has its own unique effect on a complex input signal. Except for the pure sine wave, all ac (fluctuating) signals are made up of multiple frequency components. The lowest frequency is usually the strongest in amplitude and controls the apparent frequency of the waveform as a whole. This base frequency is called the fundamental. Additional frequency components, called overtones, appear at higher frequencies than the fundamental. In a very few special cases, there may be some frequency components with frequencies lower than the fundamental. Such frequency components are called undertones.

In a periodic (repeating) waveform, the fundamental and all the overtones are harmonically related. This means that each overtone frequency is an exact, whole number multiple of the fundamental frequency. The second harmonic, for example, is two times the fundamental frequency, the third harmonic is three times the fundamental frequency, the fourth harmonic is four times the fundamental frequency, and so forth.

Not all possible harmonics are included in all waveforms. For instance, the common square wave, illustrated in Fig. 2-18, is made up of the fundamental and all odd harmonics, but no even harmonics. The following frequency components are included in a square wave:

- Fundamental,
- Third harmonic,
- Fifth harmonic,
- Seventh harmonic,
- Ninth harmonic, and
- Eleventh harmonic, etc.

The difference between various periodic waveforms lies in the harmonic content—which harmonics are included and which are omitted and the relative strength (or amplitude) of each harmonic with respect to the fundamental.

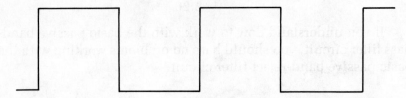

Fig. 2-18 *A square wave is made up of the fundamental and all odd harmonics, but no even harmonics.*

For the following examples, I use a sawtooth wave, because all of the harmonics are present at fairly strong amplitudes. Table 2-1 lists the harmonic content of a 250-Hz sawtooth wave. As the harmonics increase in frequency, they get weaker, so I am ignoring everything above the fifteenth harmonic, even though the harmonic series continues above this point.

For simplicity in the following examples, assume that all filters used are ideal with infinitely sharp cutoff slopes. Each frequency component is assumed to be either fully passed or fully blocked.

Table 2-2 shows what happens when you pass the 250-Hz sawtooth wave through a low-pass filter with a cutoff frequency of 1 600 Hz. Notice that everything above the sixth harmonic is filtered out of the output signal.

Table 2-3 shows what happens when we replace the low-pass filter with a high-pass filter with the same cutoff frequency—1 600

Table 2-1 Harmonic content of a 250-Hz sawtooth wave.

Harmonic	Frequency (Hz)
Fundamental	250
Second	500
Third	750
Fourth	1 000
Fifth	1 250
Sixth	1 500
Seventh	1 750
Eighth	2 000
Ninth	2 250
Tenth	2 500
Eleventh	2 750
Twelfth	3 000
Thirteenth	3 250
Fourteenth	3 500
Fifteenth	3 750

Table 2-2 Effect of a 1 600-Hz low-pass filter on a 250-Hz sawtooth wave.

Harmonic	Input (Hz)	Output (Hz)
Fundamental	250	250
Second	500	500
Third	750	750
Fourth	1 000	1 000
Fifth	1 250	1 250
Sixth	1 500	1 500
Seventh	1 750	—
Eighth	2 000	—
Ninth	2 250	—
Tenth	2 500	—
Eleventh	2 750	—
Twelfth	3 000	—
Thirteenth	3 250	—
Fourteenth	3 500	—
Fifteenth	3 750	—

Hz. Notice that all frequency components blocked in Table 2-2 are passed in Table 2-3 and vice versa. Because the high-pass filter removes the original fundamental, the lowest and strongest remaining frequency component becomes the new apparent fundamental. This has a drastic effect on the waveshape. Not all of the remaining harmonics are harmonically related to this new apparent fundamental. In this example, the old seventh harmonic

Table 2-3
Effect of a 1 600-Hz high-pass filter on a 250-Hz sawtooth wave.

Harmonic	Input (Hz)	Output (Hz)	
Fundamental	250	—	
Second	500	—	
Third	750	—	
Fourth	1 000	—	
Fifth	1 250	—	
Sixth	1 500	—	
Seventh	1 750	1 750	(New apparent fundamental)
Eighth	2 000	2 000	
Ninth	2 250	2 250	
Tenth	2 500	2 500	
Eleventh	2 750	2 750	
Twelfth	3 000	3 000	
Thirteenth	3 250	3 250	
Fourteenth	3 500	3 500	
Fifteenth	3 750	3 750	

becomes the new apparent fundamental. The old fourteenth harmonic is the second harmonic now. Old harmonics eight through thirteen are nonharmonic (or enharmonic) overtones.

The effects of two typical band-pass filters on the 250-Hz sawtooth wave are illustrated in Tables 2-4 and 2-5. Once again, the original fundamental is deleted from the output signal. The lowest and strongest remaining harmonic becomes the new apparent fun-

Table 2-4 Effect of a typical band-pass filter on a 250-Hz sawtooth wave ($F_c = 2\ 100$ Hz, BW = 800 Hz).

Harmonic	Input (Hz)	Output (Hz)
Fundamental	250	—
Second	500	—
Third	750	—
Fourth	1 000	—
Fifth	1 250	—
Sixth	1 500	—
Seventh	1 750	1 750
Eighth	2 000	2 000
Ninth	2 250	2 250
Tenth	2 500	2 500
Eleventh	2 750	—
Twelfth	3 000	—
Thirteenth	3 250	—
Fourteenth	3 500	—
Fifteenth	3 750	—

Table 2-5 Effect of a second typical band-pass filter on a 250-Hz sawtooth wave ($F_c = 2\,200$ Hz, BW = $2\,000$ Hz).

Harmonic	Input (Hz)	Output (Hz)
Fundamental	250	—
Second	500	—
Third	750	—
Fourth	1 000	—
Fifth	1 250	1 250
Sixth	1 500	1 500
Seventh	1 750	1 750
Eighth	2 000	2 000
Ninth	2 250	2 250
Tenth	2 500	2 500
Eleventh	2 750	2 750
Twelfth	3 000	3 000
Thirteenth	3 250	—
Fourteenth	3 500	—
Fifteenth	3 750	—

Table 2-6 Effect of a typical band-reject filter on a 250-Hz sawtooth wave ($F_c = 2\,100$ Hz, BW = 800 Hz).

Harmonic	Input (Hz)	Output (Hz)
Fundamental	250	250
Second	500	500
Third	750	750
Fourth	1 000	1 000
Fifth	1 250	1 250
Sixth	1 500	1 500
Seventh	1 750	—
Eighth	2 000	—
Ninth	2 250	—
Tenth	2 500	—
Eleventh	2 750	2 750
Twelfth	3 000	3 000
Thirteenth	3 250	3 250
Fourteenth	3 500	3 500
Fifteenth	3 750	3 750

damental, disrupting the harmonic relationship of the remaining overtones.

Finally, the 250-Hz sawtooth wave is fed through a typical band-reject filter in Table 2-6. Notice that the fundamental and the lowest harmonics along with the highest harmonics are left alone. Just a few of the middle harmonics are removed from the output signal by the filter.

❖ 3
Active low-pass filters

SO FAR WE HAVE DEALT ONLY WITH PASSIVE FILTER CIRCUITS. A passive circuit has no power source of its own. It "steals" its operating power from the signal being fed through it. This means a passive circuit cannot amplify or boost the signal. Passive filter circuits offer the advantages of being very simple and inexpensive, requiring just two or three simple components. By definition, you never have any power supply problems.

However, passive filter circuits also suffer from some very serious disadvantages. They are terribly imprecise and have a very gradual cutoff slope, so they can only provide a fair amount of filtering. Because they use parasitic power (operating power is taken from the signal being filtered), the entire input signal is significantly attenuated. This is called insertion loss, because some of the signal is lost simply by inserting it (feeding it) into the passive filter circuit.

Because of these inherent limitations, passive filter circuits are not used very often. For the most part, passive filter networks are used only in "quick and dirty" applications where circuit size, cost, and simplicity are the primary concerns.

The insertion loss problem can be largely overcome by placing an amplifier circuit in series with the passive filter circuit. The amplifier stage can be placed either before the filter stage, as shown in Fig. 3-1, or after the filter stage, as shown in Fig. 3-2.

In Fig. 3-1, the entire input signal is amplified or boosted before it is fed into the filter network. This precompensates for the attenuation due to the filter's insertion loss. It might seem like a waste of amplifier power to amplify the frequency components that are to be filtered out of the signal, but generally, the difference in power consumption will be negligible.

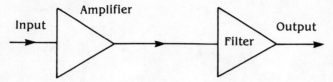

Fig. 3-1 *In an active filter, the amplifier stage can be placed before the filter stage.*

The alternate approach, shown in Fig. 3-2, boosts the signal after it has been filtered. This is usually the less desirable approach because the attenuated output signal from the filter stage may be weakened enough for any noise pickup to make a substantial contribution to the overall amplitude of the signal as a whole. Of course, neither Fig. 3-1 nor Fig. 3-2 can do anything to improve the shallow cutoff slope of the passive filter circuit.

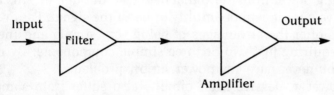

Fig. 3-2 *Alternatively, an active filter can be created by placing the amplifier after the filter stage.*

In practical electronics work, most serious filtering is done by active filter circuits. An active filter circuit incorporates its own built-in amplifier within the filter circuitry. In some cases, an active filter circuit can turn out to be less expensive and more compact than a comparable passive filter circuit, especially for band-pass or band-reject filters operating at low frequencies. This is because most active filter circuits do not require bulky and expensive inductors.

Early circuits were built around vacuum tubes, but today virtually all active filter circuits are designed around semiconductors, usually a type of IC known as an operational amplifier (op amp).

Basics of operational amplifiers

I do not have the space in this book to go into an in-depth study of the operational amplifier. A number of fine books solely on op amp theory and circuits are available. However, it might be useful to offer a brief review of the subject here. If you already know the basics of op amp circuits, you might want to skim quickly through this section.

Operational amplifiers were originally designed to perform various mathematical operations in analog circuits. In a way, an op amp is a sort of simple, nonprogrammable analog computer. Op amp circuits require a great many components. An op amp circuit built from discrete components would be bulky and very expensive. For this reason, op amps weren't used very much until they were available in inexpensive and convenient IC form.

A modern op amp IC can cost from less than a dollar up to a few dollars and will sit comfortably on your thumb. They are so inexpensive and easy to use that countless applications have been found for this once exotic part, making it perhaps the most common circuit element in modern electronics.

The de facto standard for op amps is the classic 741 chip. At one time, the 741 was the very height of high technology and was considered a very high-performance device. Today the 741 is relegated to noncritical applications and general experimentation. The technology has improved greatly since this chip was designed.

The 741 isn't always the best choice for critical filter circuits because it is relatively noisy (compared to some newer, improved op amp chips); that is, the internal circuitry of the op amp itself can add some noise (unwanted, random frequency components) to the signal passing through it. But even though it is a "low-grade" op amp by modern standards, the 741 still works very well and there continues to be a healthy market for this chip, especially because it typically sells for about 50 cents or less.

Most other op amp ICs are designed to be pin-for-pin compatible with the 741; that is, each pin serves the same function on the 741 and other op amp chips. This way, you can design and experiment with a circuit using a cheap 741. If the IC gets damaged during the course of your experimentation, it's no great loss. Once the circuit has been finalized, the 741 can be replaced with a higher grade op amp IC. An IC socket is almost essential for this sort of planned substitution.

Op amp ICs such as the 741 are offered in a variety of packaging styles. The 8-pin DIP housing, shown in Fig. 3-3, is probably the most commonly used. Sometimes a 14-pin DIP housing is

58 Active low-pass filters

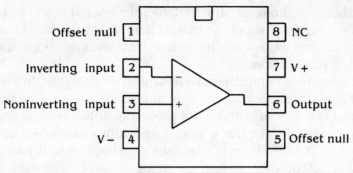

Fig. 3-3 *The popular 741 op amp IC is often supplied in an 8-pin DIP housing.*

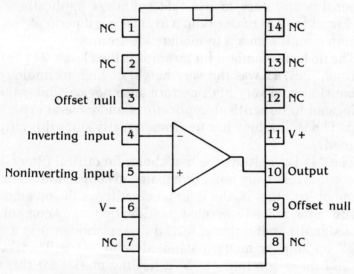

Fig. 3-4 *Sometimes a 14-pin DIP housing is used to house a 741 op amp IC.*

used, as illustrated in Fig. 3-4. Some op amp ICs are in 10-pin round cans, as shown in Fig. 3-5.

Most op amps require a symmetrical dual-polarity power supply; that is, both a positive and a negative supply voltage must be applied to the op amp. These voltages must be symmetrical around ground potential (0 V). The positive voltage should be equal to the negative voltage, except for the reversed polarity.

The supply voltages for an op amp are sometimes called supply rails. Only the actual supply rails (the positive supply voltage—V + — and the negative supply voltage—V −) are connected directly to the op amp IC. No direct ground connection is made

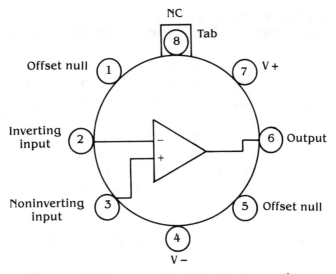

Fig. 3-5 *Some op amp ICs are in 10-pin round cans.*

to the chip, but the ground point is normally used with both the input and the output signals. A few recently developed op amp ICs are specially designed to be operated off of a single-polarity supply voltage.

The standard symbol for an op amp is shown in Fig. 3-6. Notice that there are two signal inputs and one signal output. The two inputs are marked "+" and "−." The "−" input is called the inverting input. Any signal applied to this input will have its polarity reversed or inverted at the output. If the input

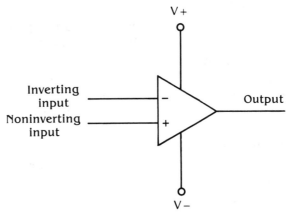

Fig. 3-6 *The standard symbol for an op amp. Power supply connections often aren't shown in schematics.*

signal is positive, the output signal will be negative and vice versa. An ac input signal will effectively be phase shifted 180°.

The other input to an op amp is marked " + ." This is the noninverting input. The polarity of a signal fed to this input is not reversed or inverted at the output. If the input signal is positive, the output signal will also be positive. If the input signal is negative, the output signal will be negative too. An ac signal fed to the inverting input of an op amp is not phase shifted.

With these two opposing inputs, an op amp is technically known as a differential amplifier. The output is equal to the inverting input voltage minus the noninverting input multiplied by the amplifier's gain.

In many practical applications, only one of the op amp's two inputs is used. The unused input is normally shorted to ground effectively making its voltage zero.

By itself, an op amp has an extremely high gain. Theoretically this open-loop gain is infinite. Of course, no practical component can ever achieve true infinite gain. In practice, the gain is just very, very high. For a 741 op amp the open-loop gain is 200 000. Some higher grade op amps have even higher open-loop gains.

For most applications, this is just too much gain. The output voltage can't exceed the supply voltages, so even a small input signal can saturate the output resulting in clipping distortion. In most practical circuits using op amps, the gain is reduced with negative feedback. Some of the output signal is fed back (through a resistance) to the inverting input. This feedback signal is subtracted from the total output signal and the amplifier's gain is effectively reduced.

Figure 3-7 shows one of the most basic op amp circuits. This is an inverting amplifier circuit. Only the inverting input is used. The noninverting input is shorted to ground, often through a resistor. The polarity of the output signal from this circuit is always the opposite of the polarity of the input signal. If the input signal is an ac waveform, it is effectively phase shifted 180°.

Two resistors determine the gain of an inverting amplifier circuit. These are the input resistor (R_i) and the feedback resistor (R_f). The ratio of these two resistances defines the circuit gain according to the simple formula

$$G = \frac{-R_f}{R_i}$$

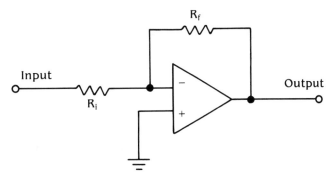

Fig. 3-7 *The inverting amplifier is one of the most basic op amp circuits.*

The minus sign in this equation indicates the polarity inversion of the signal between the input and the output.

If the feedback resistance (R_f) is large in comparison with the input resistance (R_i), the amplifier's gain will be fairly large. On the other hand, if R_f is significantly smaller than R_i, the gain will be very low. In fact, the signal will be attenuated rather than amplified in this case.

If the input resistor and the feedback resistor are equal, the circuit will exhibit unity gain. The output amplitude will equal the input amplitude. Of course, the output signal will be polarity inverted with respect to the input signal. This works even if both resistors are made equal to 0 Ω; that is, the actual resistors can be eliminated from the circuit altogether, as shown in Fig. 3-8. This circuit is known as an inverting voltage follower.

If the noninverting input is used instead of the inverting input, as shown in Fig. 3-9, you have a noninverting amplifier circuit. The output signal will have the same polarity as the input

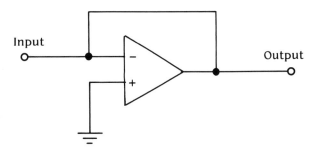

Fig. 3-8 *If the resistors are eliminated from the inverting amplifier circuit of Fig. 3-7, the result is an inverting voltage follower.*

Active low-pass filters

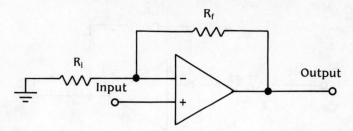

Fig. 3-9 *Another basic op amp circuit is the noninverting amplifier.*

signal. The gain formula for a noninverting amplifier is

$$G = \frac{1 + R_f}{R_i}$$

Notice that the gain can never drop below 1 (unity) in this type of circuit.

Many practical op amp circuits use both the inverting input and the noninverting input. Often either the input resistor or the feedback resistor (or perhaps both) will be replaced with some other component, such as a capacitor or a transistor. Of course, this affects the operation of the op amp. Using a capacitor, for example, will give the op amp a definite nonlinear frequency response.

The op amp has literally thousands of applications in modern electronics and there is no way I can even begin to discuss them all here. Of course, the op amp application this book is concerned with is its use in active filter circuits.

The Butterworth low-pass filter

A number of different types of active filter circuits include the Butterworth filter and the Chebyshev filter. These names might sound unduly intimidating until you realize that these circuits were simply named after the men who developed them.

This chapter looks at a few of the more common and important active low-pass filter circuits, beginning with the Butterworth filter. The math involved in the design of active filter circuits is more complex than the simple equations required for working with passive filter circuits. This complexity is simply "the nature of the beast." I can only try to make it as painless as possible.

Before you consider Butterworth filtering, you need to take a moment to consider filter orders. The order of a filter identifies its cutoff slope. The higher the order number, the steeper the cutoff slope is. Filter orders increase in steps of 6 dB/octave.

A single-stage passive filter has a cutoff slope of 6 dB/octave; that is, each doubling of frequency (octave) past the nominal cutoff frequency represents a further reduction of 6 dB in the signal level. All single-stage passive filters are first-order filters.

The simplest active filter circuits are also first-order filters with a cutoff slope of 6 dB/octave. More advanced filter circuits might have higher orders, in multiples of 6 dB/octave. For example, a second-order filter has a cutoff slope of 12 dB/octave, while a third-order filter has a cutoff slope of 18 dB/octave.

A Butterworth filter is designed to have a very flat frequency response within its passband and then a smooth, uniform roll-off characteristic. Figure 3-10 shows a frequency response graph for a typical first-order Butterworth filter. Of course, higher order Butterworth filters have steeper roll-offs. The multiple frequency response graph shown in Fig. 3-11 compares six Butterworth filters ranging from first order to sixth order. All six of these filters are assumed to have the same cutoff frequency. Notice that above the cutoff frequency, any individual frequency component experiences greater attenuation for higher filter orders. The Butterworth filtering response is considered to be more or less the "standard" filtering response.

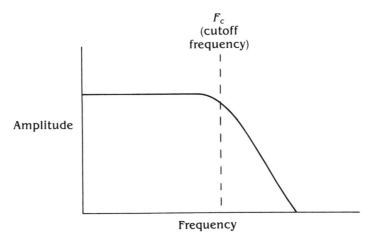

Fig. 3-10 *A frequency response graph for a typical first-order Butterworth filter.*

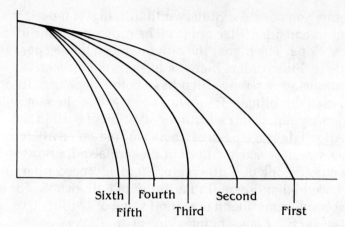

Fig. 3-11 *This multiple frequency response graph compares six Butterworth filters ranging from first order to sixth order.*

The simple passive filter networks described in chapters 1 and 2 are basically Butterworth filters. A very simple active low-pass Butterworth filter circuit is shown in Fig. 3-12. I get into specific design details of practical active low-pass filter circuits later in this chapter.

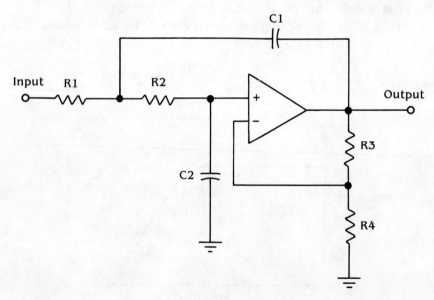

Fig. 3-12 *A very simple active low-pass Butterworth filter circuit.*

The Chebyshev low-pass filter

Another common type of filter response is exhibited by the Chebyshev filter. A frequency response graph for a typical low-pass Chebyshev filter is shown in Fig. 3-13. Notice that the frequency response is not flat below the cutoff frequency, as was the case with the Butterworth filter. The Chebyshev filter has a slight dip in amplitude below the cutoff frequency, then the amplitude goes back up to the flat level just before the actual cutoff slope begins. The chief advantage of the Chebyshev filter is its very steep roll-off characteristic.

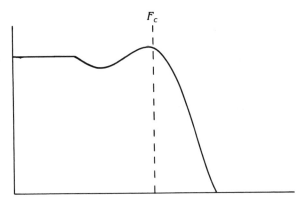

Fig. 3-13 *A frequency response graph for a typical Chebyshev filter.*

Actual circuits for Butterworth filters and Chebyshev filters are usually quite similar. Often, the only real difference in the two filter types is in the actual component values used. Both Butterworth and Chebyshev filters can be of the high-pass, band-pass, or band-reject type, as well as the low-pass type shown here.

In some sources you might find Chebyshev spelled differently. Some typical spellings include Tschebyscheff or Tschebyshev. These variations are due to the fact that this is a Russian name, and the Russian language uses a different alphabet, so the "correct" English spelling is somewhat a judgment call. Don't let such alternate spellings confuse you. If it looks like the pronunciation is reasonably close to Chebyshev, you can probably safely assume it's just an alternate spelling and not a new term.

Infinite gain multiple feedback low-pass filter circuits

Now we can finally move on to some actual active low-pass filter circuits. One of the simplest types of active filter is the infinite gain multiple feedback filter. This type of circuit is built around an op amp operating in its (theoretical) infinite gain mode with several feedback paths. Figure 3-14 shows an infinite gain multiple feedback low-pass filter circuit. Five passive components (three resistors and two capacitors) are used in addition to the op amp.

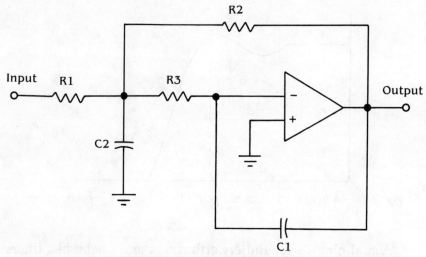

Fig. 3-14 *An infinite gain multiple feedback low-pass filter circuit.*

The design equations are rather complicated, so I make some assumptions and take some shortcuts. The first step in designing an infinite gain multiple feedback low-pass filter circuit is to calculate a variable called G_2. The G_2 value will vary with the filter order, the amplifier gain, and the normalized value of C_1. The normalized value of C_1 is an arbitrary value used for calculations. It is not a final component value. A good normalized value for C_1 is 0.01 μF. Capacitor C2 is always given a normalized value of 1 μF to prevent the calculations from being too complex.

The Butterworth version

Table 3-1 lists some precalculated values for G_2 for second through fifth order filters with gains of 5, 10, 15, or 20. A But-

Table 3-1 Typical values of G_2 for designing infinite gain multiple feedback low-pass filter circuits.

Order	Gain			
	5	10	15	20
2	0.208 2	0.109 1	0.071 7	0.051 7
3	0.277 2	0.147 7	0.098 9	0.073 3
4	0.346 1	0.185 8	0.125 6	0.093 9
5	0.413 9	0.223 1	0.151 5	0.113 9
6	0.480 6	0.259 7	0.176 8	0.133 4
7	0.546 4	0.295 8	0.201 8	0.152 5

terworth filter pattern is assumed here. Chebyshev filters would give different values. This table will save you several steps of calculation. Later on, I break down the required equation so you won't be limited to the assumptions I have made in drawing up this table. This table is presented as a convenience for readers who prefer to avoid as much math as possible when designing their circuits.

These G_2 values are measures of conductance, the reciprocal of resistance, measured in mhos. To convert between ohms and mhos, take the reciprocal of the specified value:

$$\text{ohms} = \frac{1}{\text{mhos}}$$

$$\text{mhos} = \frac{1}{\text{ohms}}$$

Once you've selected the desired gain and order for our infinite gain multiple feedback filter and have found the appropriate value of G_2, what do you do next? The next step is to find a couple of additional conductance values: G_1 and G_3. Values G_1 and G_2 determine the filter's amplifier gain:

$$K = \frac{G_1}{G_2}$$

where
  K = gain.

At this point, you know the values of G_2 and K, so you can easily rearrange this equation to solve for G_1:

$$G_1 = KG_2$$

The formula for the third conductance value (G_3) is

$$G_3 = \frac{C_1 b_0}{G_2}$$

For a Butterworth filter, the value of b_0 is always one. (This will not always be the case for Chebyshev filters.) The value C_1 is the normalized value of capacitor C1. I strongly recommend sticking with a standard value of 0.01 µF. Because C_1 and b_0 are effectively constants (for Butterworth filters), you can simplify the G_3 equation to

$$G_3 = \frac{0.01 \times 1}{G_2}$$

$$= \frac{0.01}{G_2}$$

Notice that so far I have not even mentioned the desired cutoff frequency of the filter circuit you are designing. This important factor comes into play as you start to denormalize the required resistor and capacitor values to find the specific component values needed to actually build the final circuit. First, you need to use a couple more intermediate equations to derive two normalizing factors: u and ISF.

The value u is the frequency normalizing factor and its formula is:

$$u = 2\pi F_c$$

where
 π = 3.14, and
 F_c = desired cutoff frequency for the filter circuit being designed.

Because pi is a constant, this formula can be rewritten as

$$u = 6.28 F_c$$

ISF is the impedance scaling factor and the formula for this factor is

$$ISF = \frac{F_c}{20\pi}$$

$$= \frac{F_c}{62.8}$$

Now you can use these factors to denormalize the actual component values. For the two capacitors, the denormalizing equation is

$$C_a = \frac{C_n}{(u)(ISF)}$$

where
  C_a = actual (denormalized) capacitor value, and
  C_n = normalized capacitor value.

By standardizing the normalized capacitor values ($C_1 = 0.01$ μF and $C_2 = 1$ μF), you can be more specific in your equations:

$$C_1 = \frac{0.01}{(u)(ISF)}$$

$$C_2 = \frac{1}{(u)(ISF)}$$

The formulas for the three resistor values in the circuit are

$$R_1 = \frac{ISF}{G_1}$$

$$R_2 = \frac{ISF}{G_2}$$

$$R_3 = \frac{ISF}{G_3}$$

In other words, to find the appropriate resistances, take the reciprocals of the three conductance values (derived earlier) and multiply each by the impedance scaling factor.

All these equations probably seem a little confusing and perhaps even a bit intimidating. Let's go step by step through a typical design example. The goal in this example is to design a Butterworth-type infinite gain multiple feedback second-order low-pass filter circuit with a cutoff frequency of 5 000 Hz and a gain of 10.

The first step is to look up the appropriate value for G_2 in Table 3-1. In this case, G_2 equals 0.109 1. Next, solve for G_1:

$$G_1 = KG_2$$
$$= 10 \times 0.109\ 1$$
$$= 1.091$$

70   Active low-pass filters

Then solve for G_3:

$$G_3 = \frac{0.01}{G_2}$$

$$= \frac{0.01}{0.109\ 1}$$

$$= 0.091\ 7$$

The next step is to find the frequency normalizing factor,

$$u = 6.28 F_c$$
$$= 6.28 \times 5\ 000$$
$$= 31\ 400$$

and then the impedance scaling factor:

$$\text{ISF} = \frac{F_c}{62.8}$$

$$= \frac{5\ 000}{62.8}$$

$$= 79.617\ 8$$

Now you have all of our intermediate factors, and you can begin to find the actual, denormalized component values to use in your finished sample circuit. First, solve for the capacitor values:

$$C_1 = \frac{0.01}{u \times \text{ISF}}$$

$$= \frac{0.01}{31\ 400 \times 79.617\ 8}$$

$$= \frac{0.01}{25\ 000\ 000}$$

$$= 0.000\ 000\ 000\ 4 \text{ F}$$

$$= 0.000\ 4\ \mu\text{F}$$

$$C_2 = \frac{1}{u \times \text{ISF}}$$

$$= \frac{1}{31\ 400 \times 79.617\ 8}$$

$$= \frac{1}{25\ 000\ 000}$$

$$= 0.000\ 000\ 04 \text{ F}$$

$$= 0.04\ \mu\text{F}$$

Finally, solve for the three resistor values:

$$R_1 = \frac{ISF}{G_1}$$

$$= \frac{79.617\,8}{1.091}$$

$$= 73\ \Omega$$

$$R_2 = \frac{ISF}{G_2}$$

$$= \frac{79.617\,8}{0.109\,1}$$

$$= 730\ \Omega$$

$$R_3 = \frac{ISF}{G_3}$$

$$= \frac{79.617\,8}{0.091\,7}$$

$$= 868\ \Omega$$

You can probably round these calculated resistance values off to the nearest standard resistor values without throwing off the actual cutoff frequency of the finished filter circuit too much:

- $R_1 = 75\ \Omega$
- $R_2 = 750\ \Omega$
- $R_3 = 910\ \Omega$

Calculating G_2 directly

For most general design work you can probably get by with using the G_2 values from Table 3-1. For those of you with more of a mathematical inclination, or if you need more specific values, I now cover the direct calculation of G_2. If you don't like a lot of math and don't feel you will need to do such calculations directly, feel free to skip ahead to the next section of this chapter.

Actually, the math involved really isn't all that bad. The formula for calculating G_2 is

$$G_2 = \frac{b_1 + \sqrt{b_1^2 - 4C_1 b_0(1 + K)}}{2(1 + K)}$$

where
- C_1 = normalized value of capacitor C1 (0.01 µF is a recommended standard normalization value), and
- K = desired gain.

The values b_0 and b_1 are selected to suit the desired filter order. For a Butterworth filter, b_0 will always have a value of 1, so this factor can be ignored in the equation. (This is not true of Chebyshev filters, which are covered in the next subsection of this chapter.) The appropriate b_1 value can be taken from Table 3-2, according to the desired filter order.

Order	b_1
2	1.414 2
3	2.000 0
4	2.613 1
5	3.236 1
6	3.863 7
7	4.494 0

Table 3-2 Values of b_1 for Butterworth filters.

As an example, suppose you want a fourth-order filter with a gain of 15. What is the resulting value of G_2? First, check Table 3-2 to find the value of b_1. In this case b_1 equals 2.613 1. Therefore, the value of G_2 works out to

$$G_2 = \frac{b_1 + \sqrt{b_1^2 - (4C_1 b_0 (1+K))}}{2(1+K)}$$

$$= \frac{2.613\,1 + \sqrt{2.613\,1^2 - (4 \times 0.01 \times 1 \times (1+15))}}{2(1+15)}$$

$$= \frac{2.613\,1 + \sqrt{2.613\,1^2 - (0.04 \times 16)}}{(2 \times 16)}$$

$$= \frac{2.613\,1 + \sqrt{6.828\,3 - 0.64}}{32}$$

$$= \frac{2.613\,1 + \sqrt{6.188\,3}}{32}$$

$$= \frac{2.613\,1 + 2.487\,6}{32}$$

$$= \frac{5.100\,7}{32}$$

$$= 0.159\,4$$

You can double-check your work by comparing this calculated value with the one given in Table 3-1. They are very close.

The Chebyshev version

Designing a Chebyshev-type infinite gain multiple feedback low-pass filter circuit is just a little more complicated. The b_0 and b_1 values are changed and the ripple width (the size of the dip just before the cutoff frequency) must be taken into account.

Table 3-3 lists the b_0 and b_1 values for each filter order from two to seven for five common ripple widths (0.1, 0.5, 1, 2, and 3 dB). Other than the changes in the b_0 and b_1 values, the design equations for the Chebyshev version of the infinite gain multiple feedback low-pass filter circuit are the same as for the Butterworth version. Notice that you cannot use the G_2 values from Table 3-1 for designing a Chebyshev filter circuit.

VCVS low-pass filter circuits

Another popular type of active filter circuit is the VCVS (voltage-controlled voltage source) filter. In some technical literature, the VCVS filter might be called a salen and key filter. The basic VCVS low-pass filter circuit is shown in Fig. 3-15. Six external passive components (four resistors and two capacitors) are required in

Table 3-3 Typical b_0 and b_1 values for Chebyshev filters.

Ripple width (dB)	First order b_0	First order b_1	Second order b_0	Second order b_1	Third order b_0	Third order b_1	Fourth order b_0	Fourth order b_1
0.1	6.552 2	—	3.313 3	2.372 1	1.638 1	2.629 5	0.828 5	2.025 5
0.5	2.862 8	—	1.516 2	1.425 6	0.715 7	1.534 9	0.379 1	1.025 5
1.0	1.965 2	—	1.102 5	1.097 7	0.491 3	1.238 4	0.275 6	0.742 6
2.0	1.307 6	—	0.823 0	0.803 8	0.326 9	1.022 2	0.205 8	0.516 8
3.0	1.002 4	—	0.708 0	0.644 9	0.250 6	0.928 4	0.177 0	0.404 8

74 Active low-pass filters

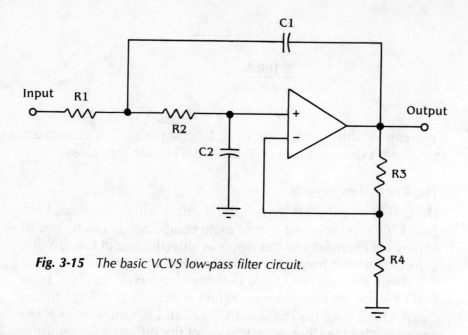

Fig. 3-15 The basic VCVS low-pass filter circuit.

addition to the op amp. The two capacitors and two of the resistors contribute to determining the cutoff frequency of the filter circuit. The formula for the cutoff frequency of the VCVS low-pass filter circuit of Fig. 3-15 is

$$F_c = \frac{1}{\left(2\pi \sqrt{R_1 R_2 C_1 C_2}\right)}$$

$$= \frac{1}{\left(6.28 \sqrt{R_1 R_2 C_1 C_2}\right)}$$

You have several possible approaches to selecting the component values for this type of circuit.

Unity gain VCVS filter

The first approach to consider is used when the passband gain of the filter is to be unity; that is, the signal level below the cutoff frequency is neither amplified nor attenuated ($K = 1$). In this system, the following equalities must hold for the component values:

- $R_1 = R_2$
- $C_1 = 2C_4$

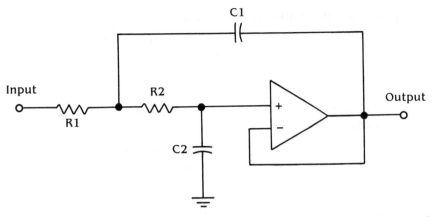

Fig. 3-16 *For a unity gain VCVS filter, resistors R3 and R4 are eliminated from the circuit.*

Resistors R3 and R4 are eliminated from the circuit in this case, leaving the filter circuit shown in Fig. 3-16.

The easiest way to avoid a lot of math is to start out with the component values for a VCVS unity gain low-pass filter with a cutoff frequency of 1 000 Hz (1 kHz). These are the standardized component values:

- $R_1 = 10$ kΩ (10 000 Ω)
- $R_2 = 10$ kΩ (10 000 Ω)
- $C_1 = 0.011\ 2\ \mu F$
- $C_2 = 0.022\ 5\ \mu F$

Now you need to scale the resistor values to account for the desired cutoff frequency. The scaling factor is equal to

$$SF = \frac{1\ 000}{F_c}$$

The two resistor values are multiplied by this scaling factor. Actually, you only have to do it once because both resistors have the same value:

$$R_n = 10\ 000 SF$$

where

R_n = new resistance value.

But this leaves you with those awkward capacitance values. So a second capacitive scaling factor is used:

$$\text{CSF} = \frac{C_n}{C_1}$$

where
C_1 = original capacitor value (0.011 2), and
C_n = new capacitor value.

Generally speaking, 0.01 μF would be a good choice throughout most of the audio frequency range, so

$$\text{CSF} = \frac{0.01}{0.011\ 2}$$

$$= 0.89$$

Because C_2 must be twice the value of C_1, you know both capacitor values:

$$C_1 = 0.01\ \mu\text{F}$$

$$C_2 = 2C_1$$
$$= 2 \times 0.01$$
$$= 0.02\ \mu\text{F}$$

The scaled resistor value (R_n) is now rescaled by dividing it by the capacitive scaling factor (CSF):

$$R = \frac{R_n}{\text{CSF}}$$

$$= \frac{10\ 000\text{SF}}{\text{CSF}}$$

If you standardize the value of C_1 to 0.01 μF, then CSF will always be equal to 0.89, so

$$R = \frac{10\ 000\text{SF}}{0.89}$$

$$= 11\ 236\text{SF}$$

$$= 11\ 236\left(\frac{1\ 000}{F_c}\right)$$

$$= \frac{11\ 236\ 000}{F_c}$$

So the actual design procedure becomes very simple. The

capacitor values are preselected and one simple calculation must be performed to give the value for the two resistors:

$$C_1 = 0.01 \ \mu F$$

$$C_2 = 0.02 \ \mu F$$

$$R_1 = R_2 = \frac{11\,236\,000}{F_c}$$

Try a couple of quick examples. If you need a cutoff frequency of 3 500 Hz (3.5 kHz), the required resistor value works out to

$$R = \frac{11\,236\,000}{3\,500}$$

$$= 3\,210 \ \Omega$$

If your intended application isn't too critical, you can round this off to the nearest standard resistance value, which happens to be 3.3 kΩ (3 300 Ω) or you can use a 2.7-kΩ (2 700-Ω) resistor in series with a 1-kΩ (1 000-Ω) trimpot to permit exact calibration.

As a second example, find the required resistance value for a VCVS low-pass filter circuit with a cutoff frequency of 800 Hz:

$$R = \frac{11\,236\,000}{800}$$

$$= 14\,045 \ \Omega$$

You could use 15-kΩ resistors or each resistance could be made up of a 12-kΩ resistor in series with a 2.2-kΩ resistor for a total of 14 200 Ω.

Remember, most resistors have a tolerance of 5% or 10%, so it rarely makes sense to knock yourself out trying to get the exact calculated value. In applications requiring high levels of precision, use a slightly smaller-valued resistor in series with a small trimpot and then calibrate the finished circuit to the exact value desired.

Because you have rescaled the capacitor values from their original normalized values, if you happen to need a filter circuit with a cutoff frequency of 1 000 Hz, the original nominal 10-kΩ (10 000-Ω) resistors won't do the trick. Instead, you would need resistors with a value of

$$R = \frac{11\,236\,000}{1\,000}$$

$$= 11\,236 \ \Omega$$

Standard 12-kΩ resistors will probably be close enough for most practical applications.

You can double-check our calculations by plugging these component values into the general cutoff frequency equation presented earlier:

$$F_c = \frac{1}{2\pi \sqrt{R_1 R_2 C_1 C_2}}$$

$$= \frac{1}{6.28 \sqrt{(12\,000 \times 12\,000 \times 0.000\,000\,01 \times 0.000\,000\,02)}}$$

$$= \frac{1}{6.28 \sqrt{0.000\,000\,028\,8}}$$

$$= \frac{1}{6.28 \times 0.000\,169\,7}$$

$$= \frac{1}{0.001\,066}$$

$$= 938 \text{ Hz}$$

That's pretty close to the targeted nominal cutoff frequency of 1 000 Hz. The difference is due to the cumulative effect of rounding off in the various equations. In a practical circuit, component tolerances are likely to cause as much error. The 12 dB/octave slope of this second-order filter also leaves some leeway in the exact cutoff frequency. If your application requires very high precision, you can use calibration trimpots in series with the two resistors in the circuit.

Equal component VCVS filter

Another easy approach to designing a VCVS low-pass filter circuit is to use the equal component method. In this system, by definition, the following component equalities must be true:

- $R_1 = R_2$
- $C_1 = C_2$

It is very important to realize that an equal component VCVS filter circuit will work only if the gain is equal to 1.586. The gain is set by the ratio of the values of resistors R3 and R4 in Fig. 3-15. In this case, the following equality must be true:

$$R_3 = 0.586 R_4$$

Good standard gain resistor values for the equal component VCVS filter circuit are

- $R_3 = 27 \text{ k}\Omega$
- $R_4 = 47 \text{ k}\Omega$

If you are using precision (1% tolerance) resistors, better results will be achieved with the following component values:

- $R_3 = 30.1 \text{ k}\Omega$
- $R_4 = 51.1 \text{ k}\Omega$

Notice that the values of resistors R3 and R4 have no effect on the cutoff frequency of the filter, so they can be standardized to the values given above for any equal component VCVS filter circuit.

Because $R_1 = R_2$ and $C_1 = C_2$, the cutoff frequency equation can be significantly simplified:

$$F_c = \frac{1}{2\pi \sqrt{R_1 R_2 C_1 C_2}}$$

$$= \frac{1}{6.28 \sqrt{R_1^2 C_1^2}}$$

$$= \frac{1}{6.28 R_1 C_1}$$

In fact, by precalculating $\frac{1}{6.28}$ and converting the capacitance value from farads to microfarads, the formula becomes

$$F_c = \frac{159\,000}{R_1 C_1}$$

Does that equation look familiar? It certainly should. It is the same equation used for the passive low-pass filter circuit discussed in chapter 1. This means that you can design an equal component VCVS filter circuit in exactly the same way as you designed our passive low-pass filter circuits earlier.

In most practical design situations, you will already know the desired cutoff frequency and will need to find the appropriate component values. This can be done by selecting a likely capacitance value and then rearranging the equation to solve for the necessary resistance value:

$$R_1 = \frac{159\,000}{F_c C_1}$$

As an example, again design a filter circuit with a cutoff frequency of 1 000 Hz (1 kHz). Use a 0.01-μF capacitor for C1 and C2. Therefore, the resistors will need to have a value of about

$$R = \frac{159\,000}{1\,000 \times 0.01}$$

$$= \frac{159\,000}{10}$$

$$= 15\,900 \; \Omega$$

A pair of 15-kΩ (15 000-Ω) resistors should be close enough.

For a second example, design an equal component VCVS low-pass filter circuit with a cutoff frequency of 5 300 Hz (5.3 kHz). This time, select 0.003 3 μF as our capacitor value, so the resistor values should be

$$R = \frac{159\,000}{5\,300 \times 0.003\,3}$$

$$= \frac{159\,000}{17.49}$$

$$= 9\,091 \; \Omega$$

A pair of 9.1-kΩ (9 100-Ω) resistors would do nicely in this circuit. If the application isn't too critical, you could even substitute common 10-kΩ (10 000-Ω) resistors.

Direct design of VCVS filter circuits

A more complex but more versatile approach to designing VCVS filter circuits is the direct method. The unity gain and equal component systems limit you to just one specific gain factor each (1 for the unity gain version or 1.586 for the equal component version). With the direct design method, your VCVS filter circuit can have almost any gain factor you desire. The circuits described so far have been strictly second-order filters. Higher order VCVS filter circuits can also be designed using the direct design method. Once again, if you're the type who prefers to avoid as much math as possible and the circuit design systems already described meet your needs, feel free to skip ahead to the next section of this chapter.

Direct design of VCVS filter circuits is very similar to the procedure used to design the infinite gain multiple feedback filter circuits in the preceding section of this chapter. Once again I use the b_0 and b_1 values from Table 3-1 (for Butterworth filters) or Table 3-3 (for Chebyshev filters). Remember that for a Butterworth filter, the value of b_0 is always 1.

The first step in the direct design of a VCVS low-pass filter circuit is to set normalized values for C_1 and C_2. I recommend $C_1 = 0.1\ \mu F$ and $C_2 = 1\ \mu F$ as good, standard normalized values for this type of circuit. Now, determine the conductance factor (G_1) using the formula:

$$G_1 = \frac{b_1 + \sqrt{b_1^2 - 4b_0(C_1 + 1 - K)}}{2}$$

where
   G_1 = conductance factor, in mhos (the reciprocal of ohms);
   K = desired gain factor; and
   C_1 = normalized value of C1, in microfarads (μF).

For a Butterworth filter, b_0 always has a value of 1, so the formula can be simplified somewhat to

$$G_1 = \frac{b_1 + \sqrt{b_1^2 - 4(C_1 + 1 - K)}}{2}$$

Using the standard normalized value of 0.1 for C_1, the equation becomes

$$G_1 = \frac{b_1 + \sqrt{b_1^2 - 4(0.1 + 1 - K)}}{2}$$

$$= \frac{b_1 + \sqrt{b_1^2 - 4(1.1 - K)}}{2}$$

Once you know G_1, you can solve for conductance G_2, using the formula

$$G_2 = \frac{C_1 b_0}{G_1}$$

Once again, for a Butterworth filter, b_0 always equals 1, so the G_2 equation can be simplified to

$$G_2 = \frac{C_1}{G_1}$$

Using the standard normalized value of 0.1 for C_1, the equation becomes

$$G_2 = \frac{0.1}{G_1}$$

So far I have not even mentioned the desired cutoff frequency of the filter circuit you are designing. This important factor comes into play as you start to denormalize the required resistor and capacitor values to find the specific component values needed to actually build the final circuit. First, you need to derive two normalizing factors: u and ISF.

The value u is the frequency normalizing factor and its formula is

$$u = 2\pi F_c$$
$$= 6.28 F_c$$

The value ISF is the impedance scaling factor and its formula is

$$\text{ISF} = \frac{F_c}{20\pi}$$
$$= \frac{F_c}{62.8}$$

Notice that these normalizing factors are the same as those used for the design of the infinite gain multiple feedback filter circuits discussed earlier in this chapter.

Now you can use these factors to denormalize the actual component values. For the two capacitors, the denormalizing equation is

$$C_a = \frac{C_n}{(u)(\text{ISF})}$$

where
C_a = actual (denormalized) capacitor value, and
C_n = normalized capacitor value.

By standardizing the normalized capacitor values ($C_1 = 0.01$ μF and $C_2 = 1$ μF), you can be more specific in your equations:

$$C_1 = \frac{0.1}{(u)(ISF)}$$

$$C_2 = \frac{1}{(u)(ISF)}$$

The formulas for the first two resistor values in the circuit are

$$R_1 = \frac{ISF}{G_1}$$

$$R_2 = \frac{ISF}{G_2}$$

In other words, to find the appropriate resistances take the reciprocals of the conductance values (derived earlier) and multiply each by the impedance scaling factor.

The only thing remaining to do is to set the values of the gain-determining resistors (R3 and R4). The formulas for these resistors are

$$R_4 = K(R_1 R_2)$$

$$R_3 = \frac{R_4}{K - 1}$$

You have a special case when the circuit has unity gain ($K = 1$). In this case, R_4 has a value of zero and R_3 is infinite (∞). In other words, both resistors can be eliminated from the circuit, as shown in Fig. 3-16. Resistor R3 is replaced by an open circuit and R4 becomes a short circuit or a direct connection.

As an example of the direct design of a VCVS low-pass filter circuit, assign the following desired specifications:

- $K = 5$
- $F_c = 6\,800$ Hz (6.8 kHz)
- Order = third
- Type = Butterworth

Because this is a Butterworth filter, b_0 is automatically set to a value of 1. For a third-order Butterworth filter with a gain of 5, b_1 equals 0.277 2.

Now you can plug these values into the G_1 equation:

$$G_1 = \frac{b_1 + \sqrt{b_1^2 - 4(1.1 - K)}}{2}$$

$$= \frac{0.2772 + \sqrt{0.2772^2 - 4(1.1 - 5)}}{2}$$

$$= \frac{0.2772 + \sqrt{0.2772^2 - 4(-3.9)}}{2}$$

$$= \frac{0.2772 + \sqrt{0.0768 + 15.6}}{2}$$

$$= \frac{0.2772 + \sqrt{15.6768}}{2}$$

$$= \frac{0.2772 + 3.9594}{2}$$

$$= \frac{4.2366}{2}$$

$$= 2.1183 \text{ mhos}$$

Next, solve for the second conductance factor (G_2):

$$G_2 = \frac{0.1}{G_1}$$

$$= \frac{0.1}{2.1183}$$

$$= 0.0472 \text{ mhos}$$

Now you're ready to set the scaling factors. First, the frequency normalization factor (u) is equal to

$$u = 6.28 F_c$$
$$= 6.28 \times 6800$$
$$= 42704$$

Then, you find that the impedance scaling factor (ISF) is equal to

$$\text{ISF} = \frac{F_c}{62.8}$$

$$= \frac{6\,800}{62.8}$$

$$= 108$$

Now you can denormalize the capacitor values:

$$C_1 = \frac{0.1}{(u)(ISF)}$$

$$= \frac{0.1}{42\,704 \times 108}$$

$$= \frac{0.1}{4\,612\,032}$$

$$= 0.000\,000\,022 \text{ F}$$

$$= 0.022\ \mu\text{F}$$

$$C_2 = \frac{1}{(u)(ISF)}$$

$$= \frac{1}{42\,704 \times 108}$$

$$= \frac{1}{4\,612\,032}$$

$$= 0.000\,000\,22 \text{ F}$$

$$= 0.22\ \mu\text{F}$$

The next step is to find the values for resistors R1 and R2:

$$R_1 = \frac{ISF}{G_1}$$

$$= \frac{108}{2.183}$$

$$= 50\ \Omega$$

$$R_2 = \frac{ISF}{G_2}$$

$$= \frac{108}{0.047\,2}$$

$$= 2\,288\ \Omega$$

Unless your particular application demands extremely high precision, you can probably get away with using the nearest standard 5% resistor values. You can use a standard 47-Ω resistor for R1 and a common 2.2-kΩ (2 200-Ω) resistor for R2.

Finally, solve for the gain resistors (R3 and R4). First, solve for R4:

$$R_4 = K(R_1 R_2)$$
$$= 5(47 \times 2\,200)$$
$$= 5 \times 103\,400$$
$$= 517\,000\ \Omega$$

A standard 470-kΩ (470 000-Ω) resistor will probably be close enough for most practical applications. Next, solve for R3:

$$R_3 = \frac{R_4}{K-1}$$
$$= \frac{470\,000}{5-1}$$
$$= \frac{470\,000}{4}$$
$$= 117\,500\ \Omega$$

I'd use a 120-kΩ (120 000-Ω) resistor for R3.

This completes the design of the 6.8-kHz third-order VCVS low-pass filter circuit with a gain of 5. The following component values are needed to build the circuit:

- C_1 = 0.022 μF
- C_2 = 0.22 μF
- R_1 = 47 Ω
- R_2 = 2.2 kΩ (2 200 Ω)
- R_3 = 120 kΩ (120 000 Ω)
- R_4 = 470 kΩ (470 000 Ω)

High-order filters

In some critical applications, very high-order filters might be required. This means the filter has a very steep cutoff slope. If you need a steeper slope than you can achieve with a single-stage filter circuit, such as those discussed throughout this chapter, you can cascade two or more filter stages with the same cutoff frequency, in series, as illustrated in Fig. 3-17.

High-order filters

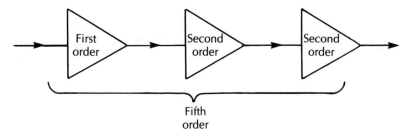

Fig. 3-17 To achieve a higher order filter, two or more filter stages can be cascaded in series.

The filter orders add. For example, two third-order filter stages and a second-order filter stage in series will exhibit a total effective order of 3 + 3 + 2 = 8. The cutoff slope will roll off at a rate of 8 × 6 = 48 dB/octave.

The overall gain of the total filter will be equal to the product of the individual filter stages included in the circuit. For example, if stage 1 has a gain of 3, stage 2 has a gain of 5, and stage 3 has a gain of 1.5, the total effective gain of the circuit as a whole will be

$$K = 3 \times 5 \times 1.5$$
$$= 22.5$$

Usually it's best to have one stage do most of the gain and set the other stages at unity gain ($K = 1$).

❖ 4
Active high-pass filters

FUNCTIONALLY A HIGH-PASS FILTER IS THE EXACT OPPOSITE OF low-pass filter. Ignoring the roll-off slope, whatever is passed by the low-pass filter is blocked by the high-pass filter and vice versa. To convert a passive low-pass filter circuit into a passive high-pass filter circuit, you had to reverse the positions of the components. (Refer to chapter 2 for more information on this). Not surprisingly, you will find that active high-pass filter circuits are remarkably similar to active low-pass filter circuits, except the positions of some components are changed. Generally speaking, the design equations used for active high-pass filter circuits are the same as those used for active low-pass filter circuits, as discussed in chapter 3.

In practical electronics work, high-pass filter circuits aren't nearly as common as low-pass filter circuits. But the high-pass filter is still extremely useful for a great many specialized applications. When passing a complex signal, such as a square wave through a low-pass filter, the original fundamental frequency is retained and the original harmonic relationships between the remaining frequency components are preserved. Just the upper harmonics are deleted from the signal. Table 4-1 illustrates the effect of a typical low-pass filter on a typical square-wave signal. The fundamental of the square-wave signal is 300 Hz and the low-pass filter has a cutoff frequency of 1 700 Hz. Ignoring the partial attenuation effects of the filter's roll-off slope, everything about the original signal's fifth harmonic (1 500 Hz) is blocked from the output signal.

Table 4-2 shows what happens when you feed that same 300-Hz square-wave signal through a high-pass filter with a cut-

Table 4-1 Effects of a low-pass filter on a typical square-wave signal.

Harmonic	Input (Hz)	Output (Hz)
Fundamental	300	300
Third	900	900
Fifth	1 500	1 500
Seventh	2 100	—
Ninth	2 700	—
Eleventh	3 300	—
Thirteenth	3 900	—
Fifteenth	4 500	—
Seventeenth	5 100	—
Nineteenth	5 700	—
Twenty-first	6 300	—
Twenty-third	6 900	—
Twenty-fifth	7 500	—

Table 4-2 Effects of a high-pass filter on a typical square-wave signal.

Harmonic	Input (Hz)	Output (Hz)	
Fundamental	300	—	
Third	900	—	
Fifth	1 500	—	
Seventh	2 100	2 100	(new apparent fundamental)
Ninth	2 700	2 700	
Eleventh	3 300	3 300	
Thirteenth	3 900	3 900	
Fifteenth	4 500	4 500	
Seventeenth	5 100	5 100	
Nineteenth	5 700	5 700	
Twenty-first	6 300	6 300	
Twenty-third	6 900	6 900	
Twenty-fifth	7 500	7 500	

off frequency of 1 700 Hz. This time the fundamental (300 Hz), as well as the third (900 Hz) and fifth harmonics (1 500 Hz), are deleted from the output signal. Everything in the original input signal above this point is passed by the filter.

Because the original fundamental frequency component has been deleted, the lowest and strongest remaining frequency component becomes the new apparent fundamental of the output signal. In this case, the new apparent fundamental is the old seventh harmonic (2 100 Hz). Notice that most of the remaining frequency components do not bear a harmonic relationship to the new apparent fundamental.

For audio signals, a high-pass filter has a much more audibly drastic effect than a low-pass filter, and even the perceived pitch of the tone will be changed because of the deletion of the original fundamental.

Strictly speaking, a true high-pass filter circuit is something of an impossibility. The circuit would theoretically have to pass everything above the filter's cutoff frequency, but no practical electronic circuit has an infinite frequency response. At some point, the circuitry will begin to act something like a low-pass filter, decreasing the signal amplitude as the frequency is increased further.

In the active high-pass filter circuits described in this chapter, this upper limit is determined by the frequency response of the op amp IC used in the circuit. This limitation will be included in the manufacturer's specifications for the chip. If your applications are typically within the audible range (below about 20 kHz (20 000 Hz), you can ignore this upper frequency limit entirely. The upper end of the flat frequency response of any standard op amp IC is well above the audible range. However, you could run into problems if you are attempting to filter high radio frequencies. You might have to use a more expensive, deluxe op amp device with an extended frequency range. As you move up through the radio frequency (RF) region, you might find you need special components that are designed for use at such high frequencies. For example, this is the case in any application involving microwaves.

Because circuits work so differently at such extremely high frequenices, I do not go into such circuitry here. The filter circuits discussed in this chapter (and the other chapters of this book) are intended for use in the audio range or at the low end of the RF spectrum.

Like the active low-pass filters discussed in chapter 3, an active high-pass filter can have either a Butterworth response (as shown in Fig. 4-1) or a Chebyshev response (as shown in Fig. 4-2). Incidentally, low-pass filters are sometimes called integrators and high-pass filters are sometimes called differentiators. These terms refer to the calculus functions that describe the action of such circuits.

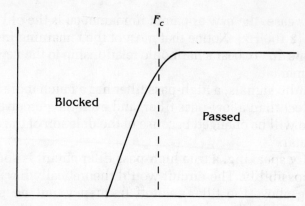

Fig. 4-1 An active high-pass filter can have a Butterworth response.

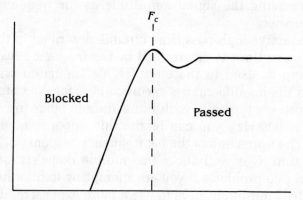

Fig. 4-2 Alternatively, an active high-pass filter can have a Chebyshev response.

Noise and active high-pass filters

For any serious filter application, it is very desirable (if not essential) to keep the filter circuit as noise-free as possible. In serious audio and other critical applications, a high-grade, low-noise op amp IC should be used. This is especially important in high-pass filter circuits, because an active high-pass filter circuit will typically tend to be noisier than a comparable active low-pass filter circuit. One reason for this is that any noise or random signal content above the useful range will be passed on to the output along with the desired signal. Transients in the signal line tend to resemble partial cycles of high-frequency signals and will be passed by a high-pass filter, whereas a low-pass filter will probably block out most (if not all) such transients.

In audio applications, high-frequency noise can often be more noticeable and objectionable than low-frequency noise at a similar amplitude. Of course, this means the output of a high-pass filter will inevitably "sound" noisier than the output from a comparable low-pass filter. You might also have a problem when an op amp chip is operated at frequencies near its upper frequency response limit. The op amp's internal circuitry can exhibit some instability at such extremely high frequencies.

The only solutions to these problems are to shield the signal lines from transient pickup as much as possible and use the best op amp IC you can. Look for one with a low noise rating. Make sure there is plenty of "headroom" between your maximum useful frequency and the upper end of the op amp's frequency response rating. Of course, this later point is unlikely to be a problem in an audio application, which by definition deals solely with frequencies well below the upper frequency response limit of even the cheapest modern op amp ICs.

Infinite gain multiple feedback high-pass filter

The infinite gain multiple feedback high-pass filter circuit is very similar in most respects to the infinite gain multiple feedback low-pass filter circuit covered in some detail in chapter 3. In the low-pass version, you worked with a pair of variables called b_0 and b_1, which are dependent on the filter order, the filter type (Butterworth or Chebyshev), and the gain. When working with high-pass filter circuits, you will be using a_0 and a_1 instead of b_0 and b_1. These new variables are simply the reciprocals of the old variables; that is,

- $a_0 = \dfrac{1}{b_0}$
- $a_1 = \dfrac{1}{b_1}$

For a Butterworth filter, you know that b_0 always has a value of 1. This means that a_0 also has a constant value of one:

$$a_0 = \frac{1}{1}$$
$$= 1$$

Remember this is true for Butterworth filters only. In a Chebyshev filter, a_0 does not normally have a value of 1. Table 4-3 lists some typical values of a_1 for Butterworth filters of various filter orders. Table 4-4 gives a_0 and a_1 values for Chebyshev filters of various filter orders with several common ripple widths.

Order	a_1
2	1.4142
3	2.0000
4	2.6131
5	3.2361
6	3.8637
7	4.4940

Table 4-3 Values of a_1 for Butterworth filters.

Table 4-4 Typical a_0 and a_1 values for Chebyshev filters.

Ripple width (dB)	First order a_0	First order a_1	Second order a_0	Second order a_1	Third order a_0	Third order a_1	Fourth order a_0	Fourth order a_1
0.1	2.3721	—	0.5283	1.2754	0.2294	0.6267	0.1280	0.3644
0.5	1.4256	—	0.3507	0.9467	0.1553	0.4243	0.0872	0.2484
1.0	1.0977	—	0.2791	0.6737	0.1244	0.3398	0.0700	0.1994
2.0	0.8038	—	0.2098	0.5064	0.0939	0.2567	0.0530	0.1509
3.0	0.6449	—	0.1703	0.4112	0.7646	0.2089	0.0432	0.1229

The schematic diagram for a second-order infinite gain multiple feedback high-pass filter circuit is shown in Fig. 4-3. Compare this with the low-pass version shown back in Fig. 3-14. Notice that the resistors and capacitors are reversed in the high-pass circuit. An infinite gain multiple feedback low-pass filter requires three resistors and two capacitors, while an infinite gain multiple feedback high-pass filter uses two resistors and three capacitors. The design procedure is simplified somewhat by the fact that capacitor C3 always has the same value as capacitor C1. This leaves just four passive components to find suitable values for when designing this type of circuit.

The design procedure can be made a lot more convenient by normalizing the value of C_1 to 1 F. Later you will denormalize this component to a more reasonable capacitance value. The normalized value for capacitor C2 is the reciprocal of the desired

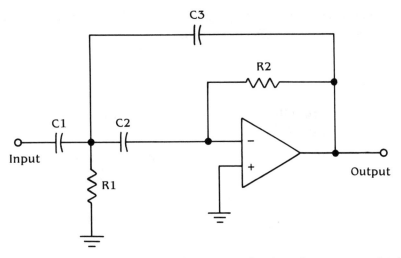

Fig. 4-3 *The schematic diagram for a second-order infinite gain multiple feedback high-pass filter circuit.*

amplifier gain for the filter circuit; that is

$$C_2 = \frac{1}{K}$$

Because you have two resistors in this circuit, you have two conductance values (G_1 and G_2) to calculate. The formulas are

$$G_1 = \frac{a_0(2K + 1)}{a_1 K}$$

$$G_2 = \frac{a_1}{2K + 1}$$

For a Butterworth filter, you know that a_0 is always 1, so the G_1 formula can be simplified to

$$G_1 = \frac{2K + 1}{a_1 K}$$

Note that this simplified version of the equation is for Butterworth filters only. This form of the equation cannot be used for the design of Chebyshev filters.

You will be using the same normalizing factors that you used with the infinite gain multiple feedback low-pass filter circuits of chapter 3. These normalizing factors are called u and ISF.

The factor u is the frequency normalizing factor and its formula is:

$$u = 2\pi F_c$$
$$= 6.28 F_c$$

where
F_c = desired cutoff frequency for the filter circuit being designed.

The factor ISF is the impedance scaling factor and its formula is

$$\text{ISF} = \frac{F_c}{20\pi}$$
$$= \frac{F_c}{62.8}$$

Now you can use these factors to denormalize the actual component values. For the two capacitor values, the denormalizing equation is

$$C_a = \frac{C_n}{(u)(\text{ISF})}$$

where
C_a = actual (denormalized) capacitor value, and
C_n = normalized capacitor value.

Of course, because C_1 is always normalized to 1 F, you can rewrite the normalizing equation as

$$C_1 = \frac{1}{(u)(\text{ISF})}$$

The remaining step in the design procedure is to denormalize the two resistor values in the circuit using the following equations:

$$R_1 = \frac{\text{ISF}}{G_1}$$

$$R_2 = \frac{\text{ISF}}{G_2}$$

In other words, to find the appropriate resistances, take the reciprocals of the two conductance values (derived earlier) and multiply each by the impedance scaling factor.

Now work your way through a typical example of designing an infinite gain multiple feedback high-pass filter circuit. Assume the following characteristics:

- $F_c = 3\,500$ Hz (3.5 kHz)
- Type = Butterworth
- Order = second
- $K = 5$

Because this is a Butterworth filter, a_0 must have a value of 1. You can get the appropriate value for a_1 from Table 4-3. In this case $a_1 = 1.414\,2$, so

$$C_2 = \frac{1}{K}$$

$$= \frac{1}{5}$$

$$= 0.2 \text{ F}$$

$$G_1 = \frac{a_0(2K + 1)}{a_1 K}$$

$$= \frac{1(2 \times 5 + 1)}{1.414\,2 \times 5}$$

$$= \frac{10 + 1}{7.071}$$

$$= \frac{11}{7.071}$$

$$= 1.555\,6 \text{ mhos}$$

$$G_2 = \frac{a_1}{2K + 1}$$

$$= \frac{1.414\,2}{2 \times 5 + 1}$$

$$= \frac{1.414\,2}{10 + 1}$$

$$= \frac{1.414\,2}{11}$$

$$= 0.128\,6 \text{ mho}$$

The next step is to find the normalizing factors:

$$u = 6.28 F_c$$
$$= 6.28 \times 3\,500$$
$$= 21\,980$$

$$\text{ISF} = \frac{F_c}{62.8}$$
$$= \frac{3\,500}{62.8}$$
$$= 55.73$$

For convenience, you can round off the value of ISF to 56. This will not introduce any significant error into your design calculations.

Now you can denormalize the component values:

$$C_1 = \frac{1}{(u)(\text{ISF})}$$
$$= \frac{1}{21\,980 \times 56}$$
$$= \frac{1}{1\,230\,880}$$
$$= 0.000\,000\,8 \text{ F}$$
$$= 0.8\,\mu\text{F}$$

$$R_1 = \frac{\text{ISF}}{G_1}$$
$$= \frac{56}{1.555\,6}$$
$$= 36\,\Omega$$

You can use either a 33-Ω or 39-Ω resistor for R1.

$$R_2 = \frac{\text{ISF}}{G_2}$$
$$= \frac{56}{0.128\,6}$$
$$= 435\,\Omega$$

You can use a 420-Ω resistor for R2. If you can't find a 420-Ω resistor (a standard, but uncommon value), you can probably get away with substituting a 470-Ω resistor.

VCVS high-pass filter circuits

The high-pass filtering function can also be performed by a type of circuit known as the VCVS filter. This stands for voltage-controlled voltage source filter. In some technical literature, the VCVS filter might be called a salen and key filter. The low-pass version of the VCVS filter circuit was introduced in chapter 3. Not surprisingly, the high-pass VCVS filter circuit is not all that different from the low-pass VCVS filter circuit.

The basic VCVS high-pass filter circuit is shown in Fig. 4-4. As in the low-pass version, this circuit requires six external passive components (four resistors and two capacitors) in addition to the op amp. If you compare this schematic with the low-pass circuit shown back in Fig. 3-14, you will see that the only major difference is that the two capacitors (C1 and C2) have traded places with the two input resistors (R1 and R3). The two gain setting resistors in the op amp's feedback path are not altered in the high-pass version of the VCVS filter circuit.

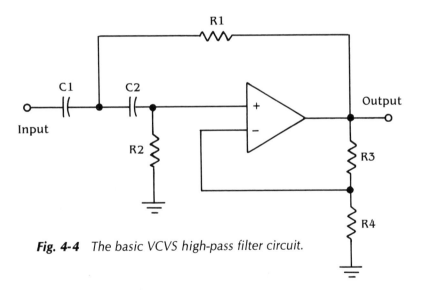

Fig. 4-4 *The basic VCVS high-pass filter circuit.*

As you probably already suspect, the design of a VCVS high-pass filter circuit isn't all that different from the design of a VCVS low-pass filter circuit. Once again, the two capacitors and two of the resistors (R1 and R2) contribute to determining the cutoff frequency of this type of filter circuit. The remaining two resistors (R3 and R4) set the amplifier gain in the circuit.

The formula for the cutoff frequency of the VCVS high-pass filter circuit of Fig. 4-4 is the same as in the low-pass version:

$$F_c = \frac{1}{2\pi\sqrt{R_1 R_2 C_1 C_2}}$$

$$= \frac{1}{6.28\sqrt{R_1 R_2 C_1 C_2}}$$

As in the case of the VCVS low-pass filter circuits discussed in chapter 3, you have several possible approaches to select the component values for the high-pass form of this circuit. Refer to chapter 3 for information on the short cut methods of VCVS filter design. Here I will concentrate on the direct method of designing the circuit.

With the direct design method, your VCVS filter circuit can have almost any gain factor you desire. This method of circuit design is also the most flexible when it comes to determining the filter order. The direct design of a VCVS high-pass filter circuit is very similar to the design methods used for the VCVS filter circuits of chapter 3 and the infinite gain multiple feedback filter circuits of both chapter 3 and the first half of this chapter.

As was the case with the infinite gain multiple feedback filter circuits from earlier in this chapter, the design of a VCVS high-pass filter circuit requires the use of the a_0 and a_1 variables, instead of the b_0 and b_1 variables used in the design of active low-pass filter circuits. Typical a_0 and a_1 values for various filter orders and gains were presented earlier in this chapter in Tables 4-3 (for Butterworth filters) and 4-4 (for Chebyshev filters).

The high-pass filter constants are almost always different from the low-pass filter constants. Remember that for a Butterworth filter, the value of a_0, like the value of b_0, is always 1.

The first step in the direct design of a VCVS high-pass filter circuit is to set normalized values for C_1 and C_2. I recommend $C_1 = C_2 = 1$ F as a good, standard normalized value for this type of circuit. You have to choose the desired filter order, cutoff frequency (F_c), and gain (K). Now you can use these known values to solve for conductance G_2 using the formula

$$G_2 = \frac{a_1 + \sqrt{a_1^2 + 8a_0(K-1)}}{4}$$

In a Butterworth filter a_0 has a constant value of 1, so this equation can be simplified to

$$G_2 = \frac{a_1 + \sqrt{a_1^2 + 8(K-1)}}{4}$$

Remember, this simplified version of the formula is valid only for Butterworth filters. For a Chebyshev filter, you must use the full form of the equation because a_0 will usually have a value other than one.

Once we know conductance G_2, we can solve for conductance G_1 using the formula

$$G_1 = \frac{a_0}{G_2}$$

Once again, since a Butterworth filter's a_0 value is always 1, you can simplify the equation to

$$G_1 = \frac{1}{G_2}$$

Remember, this simplified version of the equation is valid only for Butterworth filters. For a Chebyshev filter, you must use the full form of the equation because a_0 will usually have a value other than one.

You should also be aware that these formulas are valid only if the capacitor values (C_1 and C_2) are normalized to 1 F. If you use different normalized capacitance values, the equations will have to be changed. This makes the mathematics involved unnecessarily complicated. It's best to always use a standard normalized capacitance value of 1 F when designing a VCVS high-pass filter circuit. To build the actual working circuit, of course, you will have to denormalize the capacitor values.

For this purpose, you should use your usual normalizing factors—u (the frequency normalizing factor) and ISF (the impedance scaling factor). Use the same formulas for these factors as you have used throughout chapters 3 and 4:

$$u = 2\pi F_c$$
$$= 6.28 F_c$$

$$ISF = \frac{F_c}{20\pi}$$
$$= \frac{F_c}{62.8}$$

Then use these factors to denormalize the capacitor values:

$$C = \frac{C_n}{(u)(ISF)}$$

where
C_n = normalized capacitance value.

As I have already stated, this should always be 1 F. The equation can then be rewritten as

$$C = \frac{1}{(u)(ISF)}$$

Of course you only have to perform this equation once, because both capacitor C1 and capacitor C2 should always have the same value in this type of circuit.

The normalized value of the input resistors (R_{1n} and R_{2n}) are simply the reciprocals of the appropriate conductance values found above; that is

$$R_{1n} = \frac{1}{G_1}$$

$$R_{2n} = \frac{1}{G_2}$$

Notice that these are not the true resistance values that will be used in the actual finished circuit. These normalized resistance values are used to solve for resistors R3 and R4. The formula for R3 is

$$R_3 = \frac{ISF \times K(R_{1n} + R_{2n})}{K - 1}$$

Notice that the impedance scaling factor (ISF) is included in this equation, so it gives the final, denormalized value of R3 directly. This is also true for the formula for R4:

$$R_4 = ISF(KR_3 - R_3)$$

To denormalize the values of resistors R1 and R2, we have to multiply the appropriate normalized resistance value (R_{1n} or R_{2n}) by the impedance scaling factor (ISF):

$$R_1 = ISF \times R_{1n}$$

$$= \frac{ISF}{G_1}$$

$$R_2 = \text{ISF} \times R_{2n}$$
$$= \frac{\text{ISF}}{G_2}$$

I will now walk you through a typical design example to give you a more solid familiarity with these equations and the entire procedure. Choose the following design objectives for the example:

- $F_c = 4\,000$ Hz (4 kHz)
- Order = third
- Type = Butterworth
- $K = 10$

Of course, I will normalize the capacitance values to $C_1 = C_2 = 1$ F.

Because this is a Butterworth filter, you know a_0 has a value of 1. The appropriate value for a_1 is found from Table 4-3. In this case a_1 has a value of two.

Now you are ready to solve for conductance G_2:

$$G_2 = \frac{a_1 + \sqrt{a_1^2 + 8a_0(K-1)}}{4}$$

$$= \frac{2 + \sqrt{2^2 + 8 \times 1(10-1)}}{4}$$

$$= \frac{2 + \sqrt{4 + 8(9)}}{4}$$

$$= \frac{2 + \sqrt{4 + 72}}{4}$$

$$= \frac{2 + \sqrt{76}}{2}$$

$$= \frac{2 + 8.717\,8}{4}$$

$$= \frac{10.717\,8}{4}$$

$$= 2.679\,4 \text{ mhos}$$

The next step is to solve for conductance G_1:

$$G_1 = \frac{a_0}{G_2}$$

$$= \frac{1}{2.679\,4}$$

$$= 0.373\,2 \text{ mho}$$

The frequency normalizing factor for this filter circuit is equal to

$$u = 6.28 F_c$$
$$= 6.28 \times 4\,000$$
$$= 25\,120$$

The impedance scaling factor is equal to

$$\text{ISF} = \frac{F_c}{62.8}$$

$$= \frac{4\,000}{62.8}$$

$$= 63.694\,3$$

For convenience, you can round this off to 64.

Now we can denormalize the capacitor values:

$$C = \frac{C_n}{(u)(\text{ISF})}$$

$$= \frac{1}{25\,120 \times 64}$$

$$= \frac{1}{1\,607\,680}$$

$$= 0.000\,000\,6 \text{ F}$$

$$= 0.6 \ \mu\text{F}$$

Because C_1 and C_2 have identical values, you don't have to go through this particular equation again to solve for the second capacitor value—it's the same as the first.

The next step in our design procedure is to find the normalized values of resistors R1n and R2n:

$$R_{1n} = \frac{1}{G_1}$$

$$= \frac{1}{0.373\,2}$$

$$= 2.679\,4\ \Omega$$

Because this is just a normalized resistance value, you don't have to worry that it is ridiculously low and is not a standard resistor value. Similarly, solve for the normalized value of R2n:

$$R_{2n} = \frac{1}{G_2}$$

$$= \frac{1}{2.679\,4}$$

$$= 0.373\,2\ \Omega$$

Now you can solve for the value of resistor R3:

$$R_3 = \text{ISF} \times \frac{K(R_{1n} + R_{2n})}{K - 1}$$

$$= 64 \times \frac{10(2.679\,4 + 0.373\,2)}{10 - 1}$$

$$= 64 \times \frac{10(3.052\,6)}{9}$$

$$= 64 \times \frac{30.526}{9}$$

$$= 64 \times 3.391\,8$$

$$= 217\ \Omega$$

This is the actual resistance value to be used in the circuit. It can be rounded off to the nearest standard resistor value, which happens to be 220 Ω. This is very close in this particular case.

Next solve for the value of resistor R4:

$$R_4 = \text{ISF} \times (KR_3 - R_3)$$
$$= 64 \times (10 \times 220 - 220)$$
$$= 64 \times (2\,200 - 220)$$
$$= 64 \times 1\,980$$
$$= 126\,720\ \Omega$$

A standard 120-kΩ (120 000-Ω) resistor should be close enough.

Now all you have to do to complete the design of your VCVS high-pass filter circuit is to denormalize the resistance values for

R1 and R2. The actual, denormalized value for resistor R1 is

$$R_1 = \frac{ISF}{G_1}$$

$$= \frac{64}{0.373\,2}$$

$$= 171.5\ \Omega$$

A standard 180-Ω resistor is very close to this calculated value.

The actual denormalized value for resistor R2 is

$$R_2 = \frac{ISF}{G_2}$$

$$= \frac{64}{2.679\,4}$$

$$= 23.9\ \Omega$$

You can use either a 22-Ω or a 27-Ω resistor for R2.

This completes the design of the example VCVS Butterworth high-pass filter circuit with a cutoff frequency of 4 000 Hz (4 kHz) and a gain of 10. The following component values should be used to build this circuit:

- $C_1 = 0.62\ \mu F$
- $C_2 = 0.62\ \mu F$
- $R_1 = 180\ \Omega$
- $R_2 = 22\ \Omega$
- $R_3 = 220\ \Omega$
- $R_4 = 120\ k\Omega$ (120 000 Ω)

❖ 5
Active band-pass filters

GENERALLY SPEAKING, ACTIVE BAND-PASS FILTERS ARE SOMEWHAT more complex than active low-pass or high-pass filters. They are also more versatile and, I think, more interesting. In a sense, low-pass and high-pass filters are band-pass filters of a sort. In a low-pass filter, the lower cutoff frequency (the lower end of the passband) is at some imaginary point below 0 Hz. For a high-pass filter, the upper end of the passband is determined by the frequency response of the op amp (or other active device) used to build the filter circuit.

On the other hand, as I have already discussed in chapter 2, a band-pass filter can be created by placing a low-pass filter and a high-pass filter in series. Band-pass filters are more complex because they control more variables. This is also the reason why they are more versatile.

In working with a low-pass filter circuit, your design must account for the following variables:

- Cutoff frequency (F_c)
- Gain (K)
- Filter order (steepness of roll-off slope)

When working with a band-pass filter circuit, you still need to define the desired gain and filter order. However, instead of a single cutoff frequency, you have two cutoff frequencies to set up: the upper cutoff frequency (F_h) and the lower cutoff frequency (F_l).

More commonly, the cutoff frequencies are not addressed directly, but the design is made around two strongly related variables: center frequency (F_c) and bandwidth (BW).

More about Q

In chapter 2, I introduced the concept of Q or the "quality factor" of a filter circuit. I review and expand on that information here. While our primary interest here is in the Q of filter circuits, Q is a relevant specification for many other types of circuits. In fact, Q can be of interest when dealing with almost any type of frequency sensitive circuit or device. The specifications for Q can be given for any capacitor, inductor, or tuned circuit.

Basically Q is a figure of merit. A high Q factor indicates a low-loss, high-efficiency circuit or component. A lower Q rating means the circuit or component in question has greater loss and lower efficiency. The higher the Q factor, the lower the loss.

For a tuned circuit, such as a filter, Q is defined by the resonant (or center) frequency and the bandwidth according to the simple formula

$$Q = \frac{F_c}{BW}$$

Both the center frequency (F_c) and the bandwidth (BW) are given in hertz.

While Q is called the quality factor and is often referred to as a figure of merit, it is important not to take these names too literally. It may or may not be true that the filter circuit with the highest Q is the best choice.

For a band-pass filter circuit, a high Q basically indicates a relatively narrow passband. In some applications, this is exactly what is needed. In other applications, however, a wider passband might be preferable or even essential. In such a case, a filter circuit with a lower Q should be used. Strange as it might sound, in some instances, the circuit with the lower quality factor might be vastly superior to one with a higher quality factor.

In other words, the so-called quality factor or figure of merit is not really a measurement of actual quality or superiority in any absolute sense. It is a measurement of how sharp the tuning is; that is, how selective is the frequency response of the circuit in question? A high Q indicates strong frequency selectivity. The specific application will determine whether this is a desirable quality.

To better understand how the Q factor works, let's try a few examples. Suppose you have a band-pass filter circuit with a center frequency of 1 700 Hz and a bandwidth of 500 Hz. The Q of this filter is equal to

$$Q = \frac{1\,700}{500}$$

$$= 3.4$$

If you keep the same center frequency (1 700 Hz) but increase the bandwidth to 900 Hz, the Q changes to

$$Q = \frac{1\,700}{900}$$

$$= 1.89$$

On the other hand, if you leave the bandwidth at 500 Hz but increase the center frequency to 3 800 Hz, the Q changes to

$$Q = \frac{3\,800}{500}$$

$$= 7.6$$

Increasing the center frequency or decreasing the bandwidth increases the Q of the filter. Conversely, Q can be decreased by decreasing the center frequency or increasing the bandwidth.

As a general rule, if you need a band-pass filter circuit with a fairly high Q, the circuit should be designed for a relatively high filter order (steep roll-off slope). Otherwise it is likely that most or all of the passband will be attenuated, as illustrated in the frequency response graph of Fig. 5-1.

Infinite gain multiple feedback band-pass filters

The infinite gain multiple feedback band-pass filter circuit is similar to the infinite gain multiple feedback low-pass filter circuit of chapter 3 and the infinite gain multiple feedback high-pass filter circuit of chapter 4. The basic circuitry for the band-pass version of this type of filter circuit is illustrated in Fig. 5-2. Notice that five external passive components are required in addition to the op amp—three resistors and two capacitors.

Once again, the name comes from the fact that the op amp is operated in its infinite gain mode and multiple feedback paths are

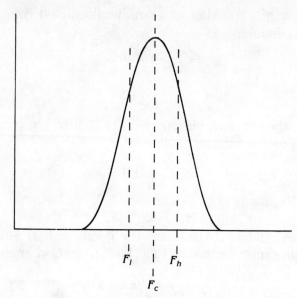

Fig. 5-1 A band-pass filter with a high Q should be of a fairly high filter order.

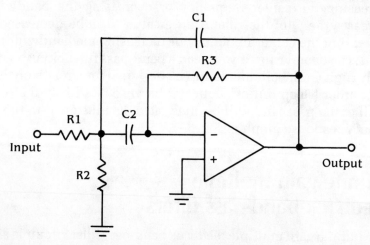

Fig. 5-2 An infinite gain multiple feedback band-pass filter circuit.

provided. The actual gain of an infinite gain multiple feedback filter circuit is not infinite. In fact, it can be quite low depending on the component values selected for use in the circuit.

This type of circuit can be designed for low to moderate gains and the Q value can be as high as about 20. Of course, lower Qs can also be selected by using the appropriate component values.

While it isn't absolutely essential, it is a very good idea to make the two capacitors in this circuit the same value ($C_1 = C_2$). This greatly simplifies the design procedure. I assume that this capacitor equality is always true as I discuss the design of the infinite gain multiple feedback band-pass filter circuit.

Before beginning the design, you must decide on the desired values for three operational values: the center frequency (F_c), the gain (K), and Q. In most applications, you will need to derive the Q value from the center frequency and the desired bandwidth of the filter's passband. Earlier, you learned that the formula for Q is

$$Q = \frac{F_c}{BW}$$

Once these variables have been selected, you can design your circuit and select the appropriate component values. The first step is to arbitrarily select a likely value for C. (Remember $C_1 = C_2 = C$.) Then it's just a matter of solving for the three resistor values using the following formulas:

$$R_1 = \frac{Q}{2\pi F_c CK}$$

$$R_2 = \frac{2Q}{2\pi F_c C}$$

$$R_3 = \frac{Q}{2\pi F_c C(2Q^2 - K)}$$

You can simplify your work a little by first solving for an intermediate value I call Y:

$$Y = 2\pi F_c C$$

This subequation appears as part of each of the resistor equations, so you only have to calculate this part once. This simplifies the resistor equations to

$$R_1 = \frac{Q}{YK}$$

$$R_2 = \frac{2Q}{Y}$$

$$R_3 = \frac{Q}{Y(2Q^2 - K)}$$

This method eliminates a little redundant work.

The gain of the circuit (K) is controlled by the values of resistors R1 and R2. The formula is

$$K = \frac{R_2}{2R_1}$$

Try out a typical example. Suppose you want to design an infinite gain multiple feedback band-pass filter circuit with a center frequency of 4 600 Hz (4.6 kHz), a bandwidth of 1 100 Hz (1.1 kHz), and a gain of 3.5. The first step is to determine the value of Q in this circuit:

$$Q = \frac{F_c}{BW}$$
$$= \frac{4\,600}{1\,100}$$
$$= 4.2$$

Next, you need to select a likely value for C. In the audio range, 0.01 μF (0.000 000 01 F) is probably a pretty good choice. Now you need to solve for our intermediate variable, Y:

$$Y = 2\pi F_c C$$
$$= 6.28 F_c C$$
$$= 6.28 \times 4\,600 \times 0.000\,000\,01$$
$$= 0.000\,288\,88$$

The remainder of the design procedure is just a matter of plugging your known values into the three resistor equations:

$$R_1 = \frac{Q}{YK}$$
$$= \frac{4.2}{0.000\,288\,88 \times 3.5}$$
$$= \frac{4.2}{0.001\,011\,1}$$
$$= 4\,154\,\Omega$$

$$R_2 = \frac{2Q}{Y}$$
$$= \frac{2 \times 4.2}{0.000\,288\,88}$$

$$= \frac{8.4}{0.000\,288\,88}$$

$$= 29\,078 \ \Omega$$

$$R_3 = \frac{Q}{Y(2Q^2 - K)}$$

$$= \frac{4.2}{0.000\,288\,88(2 \times 4.2^2 - 3.5)}$$

$$= \frac{4.2}{0.000\,288\,88(2 \times 17.64 - 3.5)}$$

$$= \frac{4.2}{0.000\,288\,88(35.28 - 3.5)}$$

$$= \frac{4.2}{0.000\,288\,88 \times 31.78}$$

$$= \frac{4.2}{0.009\,181}$$

$$= 457 \ \Omega$$

After rounding the resistances off to the nearest standard resistor values, the parts list for your example infinite gain multiple feedback band-pass filter circuit is as follows:

- $C_1 = 0.01 \ \mu F$
- $C_2 = 0.01 \ \mu F$
- $R_1 = 4.2 \ k\Omega \ (4\,200 \ \Omega)$
  $= 3.9 \ k\Omega \ (3\,900 \ \Omega)$
- $R_2 = 27 \ k\Omega \ (27\,000 \ \Omega)$
- $R_3 = 470 \ \Omega$

One important restriction to this particular circuit is that if you use a high gain (K), the Q must also be high. You cannot design this type of filter for a high gain and a low Q because the value of resistor R3 will be negative.

You might have noticed that I have not mentioned the filter order in the design of your infinite gain multiple feedback band-pass filter circuit. This is because in this circuit, the filter order is determined by Q. The higher the Q is, the steeper the roll-off slope will be. Regardless of the Q, the roll-off will approach 6 dB/octave at the extreme ends. To achieve steeper roll-off slopes (higher filter orders), you can cascade multiple stages of this circuit in series. Each stage should be set for the same center frequency, of course.

A variation on this circuit is shown in Fig. 5-3. The only difference here is that resistor R3 from the original circuit of Fig. 5-2 has been deleted. For this filter circuit to function properly, the gain (K) must be equal to twice the Q squared; that is

$$K = 2Q^2$$
$$= 2(Q \times Q)$$

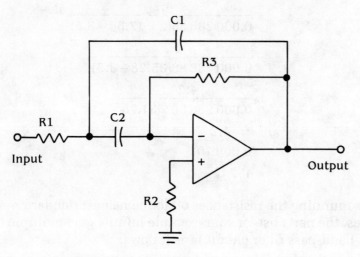

Fig. 5-3 *A variation on the basic infinite gain multiple feedback band-pass filter circuit.*

The gain of this circuit can also be derived from the two resistor values:

$$K = \frac{R_2}{2R_1}$$

Both of these equations must be simultaneously true; that is

$$K = 2Q^2 = \frac{R_2}{2R_1}$$

In addition, the two capacitors (C1 and C2) must also have equal values:

$$C_1 = C_2 = C$$

For the purposes of design, C is normalized to a convenient, standardized value of 1 F.

The conductance equations are relatively simple for this type of filter circuit:

$$G_2 = \frac{C}{2Q}$$

$$= \frac{1}{2Q}$$

$$G_1 = 2KG_2$$

To get the actual component values required to build the circuit, you must resort to your usual denormalizing factors: the frequency normalizing factor (u) and the impedance scaling factor (ISF).

$$u = 2\pi F_c$$
$$= 6.28 F_c$$

$$ISF = \frac{F_c}{20\pi}$$

$$= \frac{F_c}{62.8}$$

The formula for denormalizing the capacitor values is the same as usual too:

$$C = \frac{C_n}{(u)(ISF)}$$

where
 C_n = normalized capacitance value.

This was standardized as 1 F, so

$$C = \frac{1}{(u)(ISF)}$$

Because C1 and C2 both have the same value in this circuit, you only have to perform this calculation once.

The two resistor values are denormalized by multiplying the reciprocal of the appropriate conductance value (G_1 or G_2) by the impedance scaling factor (ISF):

$$R_1 = \frac{ISF}{G_1}$$

$$R_2 = \frac{ISF}{G_2}$$

116 Active band-pass filters

Notice that I never directly used the bandwidth (BW) in any of these design equations. The bandwidth of the filter's passband is incorporated into the Q value:

$$Q = \frac{F_c}{BW}$$

As an example of designing this type of filter circuit, suppose you need a band-pass filter with a bandwidth of 550 Hz and a center frequency of 1 400 Hz (1.4 kHz). First, you must find the Q value:

$$Q = \frac{1\,400}{550}$$
$$= 2.5$$

This means the gain of your circuit must be equal to

$$K = 2Q^2$$
$$= 2 \times 2.5^2$$
$$= 2 \times 6.25$$
$$= 12.5$$

Notice that in this case the designer has no choice over the circuit gain. It is defined directly by the required Q value.

The next step in the design procedure is to find the conductance values:

$$G_2 = \frac{C}{2Q}$$
$$= \frac{1}{2 \times 2.5}$$
$$= \frac{1}{5}$$
$$= 0.2 \text{ mho}$$

$$G_1 = 2KG_2$$
$$= 2 \times 12.5 \times 0.2$$
$$= 5 \text{ mhos}$$

Now bring in the desired center frequency ($F_c = 1\,400$ Hz) to find the denormalizing factors:

$$u = 6.28 F_c$$
$$= 6.28 \times 1\,400$$
$$= 8\,792$$

$$\text{ISF} = \frac{F_c}{62.8}$$

$$= \frac{1\,400}{62.8}$$

$$= 22.3$$

Denormalizing the capacitance value, you find that in this circuit you need two capacitors, each with a value of

$$C = \frac{C_n}{(u)(\text{ISF})}$$

$$= \frac{1}{8\,792 \times 22.3}$$

$$= \frac{1}{196\,061.6}$$

$$= 0.000\,005\,1 \text{ F}$$

Next, find the denormalized values for the two resistors in the circuit:

$$R_1 = \frac{\text{ISF}}{G_1}$$

$$= \frac{22.3}{5}$$

$$= 4.46 \; \Omega$$

$$R_2 = \frac{\text{ISF}}{G_2}$$

$$= \frac{22.3}{0.2}$$

$$= 111.5 \; \Omega$$

These resistance values are very low and aren't really too practical. But there is an easy way to fix this. You can multiply each resistance in our filter circuit by 1 000, if you divide each capacitance in the circuit by 1 000. That changes the component values to

$$C_1 = C_2 = \frac{5.1}{1\,000}$$

$$= 0.005\,1 \; \mu\text{F}$$

$$R_1 = 4.46 \times 1\,000$$
$$= 4\,460 \; \Omega$$

118 Active band-pass filters

$$R_2 = 111.5 \times 1\,000$$
$$= 111\,500\ \Omega$$

These are much more reasonable component values and will almost certainly be easier to find.

All you have to do now is round these component values off to the nearest standard values to get your completed parts list:

- $C_1 = C_2 = 0.005\ \mu F$
  $= 0.004\,7\ \mu F$
- $R_1 = 4.7\ k\Omega\ (4\,700\ \Omega)$
  $= 4.2\ k\Omega\ (4\,200\ \Omega)$
- $R_2 = 120\ k\Omega\ (120\,000\ \Omega)$

Try a second example, this time with a much higher Q. For the second example infinite gain multiple feedback band-pass filter circuit, use a center frequency of 5 800 Hz (5.8 kHz) and a bandwidth of 560 Hz. In this case, the Q is

$$Q = \frac{F_c}{BW}$$
$$= \frac{5\,800}{560}$$
$$= 10.36$$

Round this off to 10 for the sake of convenience.

With a Q this high, you can expect the gain to be very high:

$$K = 2Q^2$$
$$= 2 \times 10^2$$
$$= 2 \times 100$$
$$= 200$$

Wow! Now, that is high gain! This circuit can only be used with very low-level input signals, otherwise the output signal will be severely clipped. After all, the output voltage from an op amp can never exceed its supply voltage, even though an op amp is theoretically capable of infinite gain. But in this type of circuit, if you want a high Q, you must accept a very high gain because these two variables are so tightly interlinked. Of course, you can normalize the two capacitor values to 1 F.

Now you are ready to solve for the circuit's conductance values:

$$G_2 = \frac{C}{2Q}$$

$$= \frac{1}{2 \times 10}$$

$$= \frac{1}{20}$$

$$= 0.05 \text{ mho}$$

$$G_1 = 2KG_2$$
$$= 2 \times 200 \times 0.05$$
$$= 20 \text{ mhos}$$

The next step is to find your normalization factors:

$$u = 6.28 F_c$$
$$= 6.28 \times 5\,800$$
$$= 36\,424$$

$$\text{ISF} = \frac{F_c}{62.8}$$

$$= \frac{5\,800}{62.8}$$

$$= 92.36$$

For convenience, round off the impedance scaling factor to 92.

Next, denormalize the capacitor values. Because both capacitors have equal values, you only have to do this once:

$$C = \frac{C_n}{(u)(\text{ISF})}$$

$$= \frac{1}{36\,424 \times 92}$$

$$= \frac{1}{3\,351\,008}$$

$$= 0.000\,000\,3$$

$$= 0.3 \ \mu\text{F}$$

Some manufacturers offer 0.3-μF capacitors, but 0.33-μF capacitors are much more common and will almost certainly be close enough in all but the most extremely critical applications.

Then find the denormalized values for the two resistors in the circuit:

$$R_1 = \frac{\text{ISF}}{G_1}$$

$$= \frac{92}{20}$$
$$= 4.6 \ \Omega$$

$$R_2 = \frac{ISF}{G_2}$$
$$= \frac{92}{0.05}$$
$$= 1\,840 \ \Omega$$

Oops. Resistor R1's value is again too low to really be practical. You will find that this will often be the case for this type of circuit. In this case, multiply the resistor values by 100 and divide the capacitor values by 100, so

$$R_1 = 4.6 \times 100$$
$$= 460 \ \Omega$$

$$R_2 = 1\,840 \times 100$$
$$= 184\,000 \ \Omega$$

$$C_1 = C_2 = \frac{0.33}{100}$$
$$= 0.003\,3 \ \mu F$$

Now all you have to do is to round off the resistances to the nearest standard resistor values:

- $R_1 = 470 \ \Omega$
- $R_2 = 180 \ k\Omega \ (180\,000 \ \Omega)$

The VCVS band-pass filter

Chapter 3 looked at infinite gain multiple feedback low-pass filter circuits and VCVS low-pass filter circuits. Both these types of filter circuits were also covered in their high-pass versions in chapter 4. We have been working with the infinite gain multiple bandwidth band-pass filter circuit. Is there a VCVS band-pass filter circuit too?

There certainly is. The basic circuit for a VCVS band-pass filter circuit is shown in Fig. 5-4. This circuit is also referred to as a salen and key band-pass filter. This circuit might look a lot more complex to design because it has so many external passive

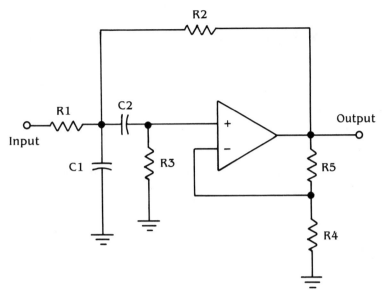

Fig. 5-4 *Another type of active band-pass filter circuit is the VCVS band-pass filter.*

components—five resistors and two capacitors—but the design procedure is significantly simplified by the fact that some of the components are always given equal values:

- $C_1 = C_2$
- $R_1 = R_2 = R_3$

You really only have to solve for four component values in designing this circuit:

- C (C_1, C_2)
- R_1 (R_1, R_2, R_3)
- R_4
- R_5

To simplify the design procedure, certain standard normalized values are first assigned to three of these components:

- $C_n = 1$ F
- $R_{1n} = 1.414$ Ω
- $R_{4n} = 1$ Ω

The first step in the design procedure is to solve for an intermediate variable I call w:

$$w = 4 - R_{1n}C_n Q$$

Because R_{1n} and C_n are always set to standard normalized values, you can simplify this equation to

$$w = 4 - (1.414 \times 1)Q$$
$$= 4 - 1.414Q$$

The same value of w can also be defined by resistances R_{4n} and R_{5n}:

$$w = 1 + \frac{R_{5n}}{R_{4n}}$$

And since the standard normalized value of R_{4n} is 1 Ω, this equation can be reduced down to

$$w = 1 + \frac{R_{5n}}{1}$$
$$= 1 + R_{5n}$$

However, at this point you don't know the value of R_{5n}, but you do know the value of w, so you can rearrange the equation to solve for R_{5n}:

$$R_{5n} = w - 1$$

So far you have dealt entirely with normalized values. The only real variable the designer has decided on is the Q of the filter, which contributes to the value of w. So now you need to call on your usual normalization factors, which are based on the filter's desired center frequency:

$$u = 6.28 F_c$$

$$\text{ISF} = \frac{F_c}{62.8}$$

The capacitance value (C) is denormalized in the usual way:

$$C = \frac{C_n}{(u)(\text{ISF})}$$

Because C_n is normalized to a standard value of 1 F, this formula can be rewritten as

$$C = \frac{1}{(u)(ISF)}$$

To denormalize the resistor values, multiply the appropriate normalized resistance value by the impedance scaling factor (ISF):

$$R_1 = ISF \times R_{1n}$$
$$R_4 = ISF \times R_{4n}$$
$$R_5 = ISF \times R_{5n}$$

Because the normalized values of R_{1n} and R_{4n} are standardized, you can rewrite the equations for R_1 and R_4 as

$$R_1 = 1.414 ISF$$
$$R_4 = 1 ISF$$
$$= ISF$$

In many (if not most) practical designs of this type, the resistor values are likely to come out too low. This can be corrected by multiplying each resistor value by a constant value and dividing the capacitor values by the same constant value. Typical multiplication constants for this purpose are 100 or 1 000.

This is actually one of the easiest active band-pass filter circuits to design, as you shall discover as you work your way through a typical example. In the example, you will design a band-pass filter with a center frequency (F_c) of 5 400 Hz (5.4 kHz), and a bandwidth (BW) of 2 900 Hz (2.9 kHz). The first step is to find the Q of the circuit:

$$Q = \frac{F_c}{BW}$$
$$= \frac{5\,400}{2\,900}$$
$$= 1.86$$

Of course, begin with your standardized normalization values for the three passive component values in the circuit:

- $C_n = 1$ F
- $R_{1n} = 1.414\ \Omega$
- $R_{4n} = 1\ \Omega$

The first step in the design procedure is to solve for w, the intermediate variable:

$$\begin{aligned} w &= 4 - 1.414Q \\ &= 4 - (1.414 \times 1.86) \\ &= 4 - 2.63 \\ &= 1.37 \end{aligned}$$

You can use w to solve for the normalized value of R_{5n}:

$$\begin{aligned} R_{5n} &= 1.37 - 1 \\ &= 0.37 \end{aligned}$$

Now you know all the normalized values in the circuit, so it is time to determine your normalization factors:

$$\begin{aligned} u &= 6.28 F_c \\ &= 6.28 \times 5\,400 \\ &= 33\,912 \end{aligned}$$

$$\begin{aligned} \text{ISF} &= \frac{F_c}{62.8} \\ &= \frac{5\,400}{62.8} \\ &= 85.99 \end{aligned}$$

You can round off the impedance scaling factor to 86. In this case, it will make no difference in the performance of the finished circuit, even in most precision applications.

Now you can normalize the capacitor value:

$$\begin{aligned} C &= \frac{1}{(u)(\text{ISF})} \\ &= \frac{1}{33\,912 \times 86} \\ &= \frac{1}{2\,916\,432} \\ &= 0.000\,000\,34 \text{ F} \\ &= 0.34 \; \mu\text{F} \end{aligned}$$

It doesn't take much rounding off to give the standard capacitor value of 0.33 μF.

The next step is to denormalize the three resistor values:

$$R_1 = \text{ISF} \times R_{1n}$$
$$= 86 \times 1.414$$
$$= 122 \text{ } \Omega$$

$$R_4 = \text{ISF} \times R_{4n}$$
$$= 86 \times 1$$
$$= 86 \text{ } \Omega$$

$$R_5 = \text{ISF} \times R_{5n}$$
$$= 86 \times 0.37$$
$$= 32 \text{ } \Omega$$

These resistor values are too low to be practical, so multiply all resistor values by 100 and divide the capacitor value by 100:

$$R_1 = 122 \times 100$$
$$= 12\,200 \text{ } \Omega$$

$$R_4 = 86 \times 100$$
$$= 8\,600 \text{ } \Omega$$

$$R_5 = 32 \times 100$$
$$= 3\,200 \text{ } \Omega$$

$$C = \frac{0.33}{100}$$
$$= 0.003\,3 \text{ } \mu F$$

All that's left to do is to round off the component values to the nearest standard values and finish the parts list for the circuit:

- C_1, C_2 = 0.003 3 μF
- R_1, R_2, R_3 = 12 kΩ (12 000 Ω)
- R_4 = 8.2 kΩ (8 200 Ω)
- R_5 = 3.3 kΩ (3 300 Ω)

It might be a good idea to place a 390-Ω resistor in series with R4 for a total of 8 590 Ω, which is very close to the desired nominal value of 8 600 Ω.

This type of VCVS band-pass filter circuit has a very serious limitation, however. It will only work for low Q values. To see why this is so, try a second example with a higher Q. This time use a center frequency of 15 000 Hz (15 kHz) and a bandwidth of 1 500 Hz (1.5 kHz). This time the Q works out to

$$Q = \frac{F_c}{BW}$$

$$= \frac{15\,000}{1\,500}$$

$$= 10$$

Now solve for the intermediate value of w:

$$w = 4 - 1.414Q$$
$$= 4 - (1.414 \times 10)$$
$$= 4 - 14.14$$
$$= -10.14$$

This means the nominal value for resistor R5n must be

$$R_{5n} = w - 1$$
$$= -10.14 - 1$$
$$= -11.14$$

The value of resistor R5 is negative, which is impossible. Clearly, there is no way to design a practical circuit that calls for negative component values.

This type of VCVS band-pass filter circuit is easy to design and works well for low Qs, but it is limited to Qs below about 2. If the Q is increased above this, the value of R5 will become negative.

High-Q VCVS band-pass filter circuits

There is no way to design a practical filter high-Q circuit of the type we've been working with using just a single op amp. But there is no law preventing you from using more than one op amp in a single circuit.

The band-pass filter circuit shown in Fig. 5-5 can be designed for Qs as high as 50. The official name for this type of circuit is a second-order band-pass positive feedback active filter. The "positive feedback" part refers to the fact that the noninverted output from the second op amp stage is fed back through resistor R3 into the input of the first output stage.

Both op amps in this circuit are operated in the inverting

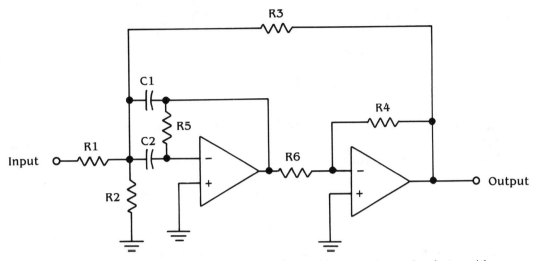

Fig. 5-5 *This variation on the basic VCVS band-pass filter circuit can be designed for very high Qs.*

mode. The first op amp stage inverts the signal, then the second op amp stage inverts the signal back to its original (noninverted) condition.

The design of this circuit is simplified somewhat by certain component equalities:

- $C_1 = C_2$
- $R_1 = R_5 = R_6$

To design this type of filter circuit, first set a couple of standardized normalization values:

- $C = 1 \text{ F } (C = C_1 = C_2)$
- $R_{3n} = 1 \text{ } \Omega$
- $R_{4n} = 1 \text{ } \Omega$

Now find the normalized values for the other two resistors in the circuit. (Remember R_5 and R_6 have the same value as R_1, so you don't have to solve for them separately.) The formula for R_{1n} is

$$R_{1n} = 2Q - \frac{QR_{4n}}{R_{3n}}$$

But because you already know that both R_{3n} and R_{4n} have been normalized to 1 Ω each:

$$R_{1n} = 2Q - \frac{(Q \times 1)}{1}$$
$$= 2Q - Q$$
$$= Q \ \Omega$$

The normalized value for resistor R1n is the same as the desired Q for the filter circuit.

Next turn to the remaining resistance value—R_{2n}:

$$R_{2n} = \frac{1}{R_{1n} - \frac{1}{R_{1n}} - \frac{1}{R_{3n}}}$$

Plugging your known normalized resistance values into this equation, you can find that R_{2n} becomes

$$R_{2n} = \frac{1}{Q - \frac{1}{Q} - \frac{1}{1}}$$
$$= \frac{1}{Q - \frac{1}{Q} - 1}$$

At this point, you have found normalized values for all of the components in the circuit. Now it's time to find the normalization factors, which are dependent on the desired center frequency (F_c):

$$u = 6.28 F_c$$

$$ISF = \frac{F_c}{62.8}$$

As usual, the formula for denormalizing the capacitor value is

$$C = \frac{C_n}{(u)(ISF)}$$
$$= \frac{1}{(u)(ISF)}$$

Of course, because both capacitors in this circuit have the same value this equation only has to be performed once.

To denormalize the resistor values, multiply each normalized resistance by the impedance scaling factor (ISF):

$$R_1 = \text{ISF} \times R_{1n}$$
$$R_2 = \text{ISF} \times R_{2n}$$
$$R_3 = \text{ISF} \times R_{3n}$$
$$R_4 = \text{ISF} \times R_{4n}$$

But resistors R3n and R4n have been normalized to the same standard value—1 Ω—so there is no need to repeat this equation. Resistor R4 will always have the same value as resistor R3. Moreover, because this resistance is normalized at 1 Ω, the formula is reduced to

$$R_3 = \text{ISF} \times 1$$
$$= \text{ISF}$$

This design procedure will probably result in some impractically low resistor values. This can be corrected by multiplying all resistor values in the circuit by a constant (usually, but not necessarily, 100 or 1 000). Any convenient value can be used, as long as it is consistent throughout the circuit. Then divide all of the capacitor values in the circuit by the same constant. This will give you more practical component values for your finished parts list and the circuit will still operate to its designed specifications.

As an example, try designing a second-order band-pass positive feedback active filter circuit with a center frequency of 4 500 Hz (4.5 kHz) and a bandwidth of 100 Hz. Obviously, this filter will have a very high Q with a very narrow passband:

$$Q = \frac{F_c}{\text{BW}}$$
$$= \frac{4\,500}{100}$$
$$= 45$$

First, you must assign a normalized value equal to Q for resistor R1n:

$$R_{1n} = Q$$
$$= 45 \text{ Ω}$$

The next step is to solve for the normalized value of resistor R2n:

$$R_{2n} = Q - \frac{1}{Q} - 1$$

$$= 45 - \frac{1}{45} - 1$$

$$= 45 - 0.022\,2 - 1$$

$$= 45 - 1.022\,2$$

$$= 43.977\,8 \, \Omega$$

You can round this off to 44 Ω. You now have all the normalized component values for the circuit:

- $C_n = 1$ F
- $R_{1n} = 45 \, \Omega$
- $R_{2n} = 44 \, \Omega$
- $R_{3n} = 1 \, \Omega = R_{4n}$

The next step in your design procedure is to find the normalizing factors:

$$u = 6.28 F_c$$
$$= 6.28 \times 4\,500$$
$$= 28\,260$$

$$\text{ISF} = \frac{F_c}{62.8}$$

$$= \frac{4\,500}{62.8}$$

$$= 71.66$$

Now denormalize the capacitor values:

$$C = \frac{C_n}{(u)(\text{ISF})}$$

$$= \frac{1}{28\,260 \times 71.66}$$

$$= \frac{1}{2\,025\,111}$$

$$= 0.000\,000\,5 \text{ F}$$
$$= 0.5 \, \mu\text{F}$$

The next step is to denormalize the three resistor values:

$$R_1 = \text{ISF} \times R_{1n}$$
$$= 71.66 \times 45$$
$$= 3\,224.7 \ \Omega$$

$$R_2 = \text{ISF} \times R_{2n}$$
$$= 71.66 \times 44$$
$$= 3\,153.04 \ \Omega$$

$$R_3 = \text{ISF} \times R_{3n}$$
$$= 71.66 \times 1$$
$$= 71.66 \ \Omega$$

This last resistor value is a bit low, so multiply all the resistor values in the circuit by 100 and divide the capacitor values by the same 100:

$$R_1 = 3\,224.7 \times 100$$
$$= 322\,470 \ \Omega$$

$$R_2 = 3\,153.04 \times 100$$
$$= 315\,304 \ \Omega$$

$$R_3 = 71.66 \times 100$$
$$= 7\,166 \ \Omega$$

$$C = \frac{0.5}{100}$$
$$= 0.005 \ \mu F$$

Finally round off these component values to the nearest standard values and complete the parts list:

- R_1, R_5, R_6 = 330 kΩ (330 000 Ω)
- R_2 = 312 kΩ (312 000 Ω) (made up of a 270-kΩ resistor in series with a 42-kΩ resistor)
- R_3, R_4 = 7.19 kΩ (7 190 Ω) (made up of a 6.8-kΩ resistor in series with a 390-Ω resistor)

High-order band-pass filters

All of the active band-pass filter circuits in this chapter have basically been second-order filters. To achieve higher filter orders, multiple filter stages can be cascaded together. In the case of band-pass filter operation, this is usually done in the form of a special type of filter circuit called a state variable filter. This type of filter circuit will be the subject of chapter 7. In the meantime, take a look at active circuitry for the fourth basic filter type—the band-reject filter.

❖ 6
Active band-reject filters

IN MANY WAYS, THE BAND-REJECT FILTER SEEMS LIKE SOMETHING of an oddball among filter circuits. Perhaps it's the name. All of the other filter types refer to what they pass, but this one only says what it doesn't pass. Perhaps it's because the frequency response graph looks rather odd. A frequency response graph for a typical band-reject filter is shown in Fig. 6-1. A band-reject filter passes almost everything except for those frequency components within a specific (usually narrow) band.

Band-reject filters are used to "edit out" specific unwanted frequency components. A typical application for a low-range band-reject filter might be to remove 60-Hz ac line noise from a low-frequency signal line. Frequency components above and below the specified reject band are passed.

As you can see from the frequency response graph of Fig. 6-1, the circuit's frequency response has a hole or notch. For this reason, the band-reject filter is often called the notch filter.

Like the band-pass filter, the operation of a band-reject filter can be defined by the lower cutoff frequency (F_l) and the upper cutoff frequency (F_h). Describing the reject band, or notch, by its center frequency and bandwidth, is more common. These terms are exactly the same as with the band-pass filter. The bandwidth is the distance between the upper cutoff frequency (F_h) and the lower cutoff frequency (F_l):

$$BW = F_h - F_l$$

Similarly, the center frequency is the exact midpoint of the reject band or notch. It can be found by taking the average of the

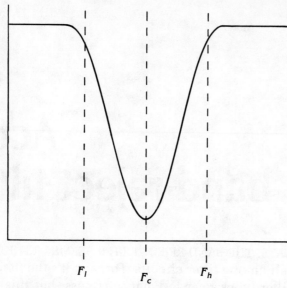

Fig. 6-1 *The frequency response graph for a typical band-reject filter.*

upper cutoff frequency (F_h) and the lower cutoff frequency (F_l):

$$F_c = \frac{F_h + F_l}{2}$$

All of these equations work in exactly the same way as for the band-pass filter except the filtering action is reversed. What is passed by the band-pass filter is blocked by the band-reject filter and vice versa.

I'll run through one quick example here. Suppose you have a band-reject filter with a lower cutoff frequency (F_l) of 330 Hz and an upper cutoff frequency (F_h) of 940 Hz. The bandwidth of this particular filter is equal to

$$\begin{aligned} BW &= F_h - F_l \\ &= 940 - 330 \\ &= 610 \text{ Hz} \end{aligned}$$

This filter's center frequency (F_c) works out to

$$\begin{aligned} F_c &= \frac{F_h + F_l}{2} \\ &= \frac{940 + 330}{2} \end{aligned}$$

$$= \frac{1\,270}{2}$$

$$= 635 \text{ Hz}$$

Band-reject filters use the Q factor just like band-pass filters, and it works in exactly the same way. That is, the Q of a filter is the ratio of its center frequency to its bandwidth:

$$Q = \frac{F_c}{BW}$$

For the example band-reject filter circuit, the Q is equal to

$$Q = \frac{635}{610}$$

$$= 1.04$$

Of course, band-reject filter circuits with higher or lower Q factors are possible. Most practical band-reject filter circuits have moderate to high Q values, indicating a very narrow bandwidth with reference to the center frequency.

Because the usual function of a band-reject filter circuit is to delete a few specific frequencies while leaving the rest of the signal alone, at least unity gain and relatively steep roll-off slopes are usually essential. This means that practical band-reject filter circuits are almost always of the active type. Passive band-reject filters are relatively rare.

The passive twin-T band-reject filter

In chapter 2, I introduced the basic passive band-reject filter circuit, which is shown again in Fig. 6-2. This is not the only type of passive band-reject filter circuit. In fact, this circuit is rarely used. When a passive band-reject filter is called for, it is more common to use a circuit called a twin-T band-reject filter circuit, shown in Fig. 6-3. In some technical literature, the name of this circuit might be spelled "twin-tee" instead of "twin-T." The reason for the name should be apparent from the schematic. The components are arranged into two T-like shapes.

The advantage of this circuit is that it doesn't require a coil. Only standard resistors and capacitors are used in the twin-T band-reject filter circuit. While there are a total of six compo-

136 Active band-reject filters

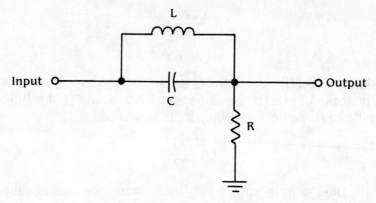

Fig. 6-2 *The basic passive band-reject filter circuit.*

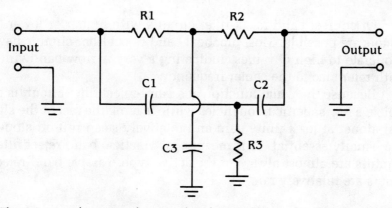

Fig. 6-3 *Another type of passive band-reject filter is the twin-T network.*

nents in this circuit, only two resistor and two capacitor values are required, because

- $R_1 = R_2$
- $C_1 = C_2$

Moreover, the second value for each component is easily derived from the first:

$$R_3 = \frac{R_1}{2}$$

$$C_3 = 2C$$

The center frequency formula for this type of circuit is

$$F_c = \frac{1}{2\pi R_1 C_1}$$

$$= \frac{1}{6.28 R_1 C_1}$$

With this type of circuit, the designer has no control over the Q, the bandwidth, or the roll-off slope. The bandwidth is moderately wide, so the Q is fairly low. And because this is a passive filter circuit, the roll-off slope is quite gradual. However, the center frequency, which is sometimes called the null frequency in this type of application, will be subjected to significant attenuation, typically about 30 to 40 dB. This is assuming, of course, that the actual component values are close to their calculated nominal values without significant rounding off.

In practical design work, you will know the desired center frequency and your task will be to find the appropriate component values. The best approach is to select a likely value for C_1 and rearrange the center frequency equation to solve for R_1:

$$R_1 = \frac{1}{6.28 F_c C_1}$$

Try a quick example. Suppose you need a passive band-reject filter circuit with a center frequency of 1 200 Hz (1.2 kHz). I will arbitrarily select a value of 0.01 µF for capacitor C1. What is the required value for resistor R1?

$$R_1 = \frac{1}{6.28 \times 1\,200 \times 0.000\,000\,01}$$

$$= \frac{1}{0.000\,075\,3}$$

$$= 13\,280 \ \Omega$$

If the application isn't too terribly critical, you can probably use a standard 5% tolerance 12-kΩ (12 000-Ω) resistor. If you need a little more precision, you can place a 1.2-kΩ (1 200-Ω) resistor in series with the 12-kΩ resistor for a total of 13.2 kΩ (13 200 Ω).

Before going on with the design, look at how far off your center frequency would be if you round off R_1 to 12 kΩ:

$$F_c = \frac{1}{6.28 \times 12\,000 \times 0.000\,000\,01}$$

$$= \frac{1}{0.0007536}$$

$$= 1\,327 \text{ Hz}$$

This may or may not be close enough depending on the specific requirements of our individual application.

But suppose that a 12-kΩ value for resistor R1 is close enough for the application in this design example. You now know four of the component values in the circuit:

- $R_1 = R_2 = 12$ kΩ (12 000 Ω)
- $C_1 = C_2 = 0.01$ μF

Now it's a simple matter to solve for the remaining component values:

$$C_3 = 2C_1$$
$$= 2 \times 0.01$$
$$= 0.02 \text{ μF}$$

$$R_3 = \frac{R_1}{2}$$
$$= \frac{12\,000}{2}$$
$$= 6\,000 \text{ Ω}$$

This is a rather tricky resistance value, but you can use a 3.3-kΩ (3 300-Ω) resistor in series with a 2.7-kΩ (2 700-Ω) resistor to get the correct total:

$$3\,300 + 2\,700 = 6\,000$$

The active twin-T band-reject filter circuit

This chapter is about active band-reject filter circuits, and Fig. 6-4 shows an active version of the twin-T circuit. Essentially, all you've done here is add an op amp stage to the output of a passive twin-T network. The center frequency is still determined by the passive components in this first half of the circuit. The added resistors (R4 and R5) and capacitor (C4) define the Q of the active twin-T band-reject filter.

The center frequency of this filter circuit is found in the

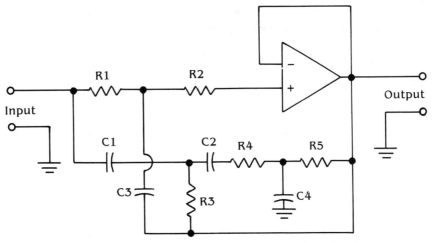

Fig. 6-4 *An active twin-T band-reject filter circuit.*

same way as in the passive version; that is,

$$F_c = \frac{1}{2\pi R_1 C_1}$$

$$= \frac{1}{6.28 R_1 C_1}$$

The following component values are dependent on the first two:

- $R_2 = R_1$
- $R_3 = \dfrac{R_1}{2}$
- $C_2 = C_1$
- $C_3 = 2C$

For this circuit to work properly, the following must be true:

$$\frac{R_4}{2R_1} = \frac{C_1}{C_4}$$

These ratios define the Q of the filter:

$$Q = \frac{R_4}{2R_1} = \frac{C_1}{C_4}$$

Assuming you already know the desired Q and the values of

resistor R1 and capacitor C1, you can rearrange these equations to solve for the unknown component values:

$$R_4 = 2R_1 Q$$

$$C_4 = \frac{C_1}{Q}$$

Finally, the remaining component (resistor R5) is always given the same value as R_4:

$$R_5 = R_4$$

Try these equations with a typical example. Design an active twin-T band-reject filter circuit with a center frequency (or null frequency) of 880 Hz and a bandwidth of 200 Hz. The Q of this circuit must be

$$Q = \frac{F_c}{BW}$$

$$= \frac{880}{200}$$

$$= 4.4$$

The first step in designing this filter circuit is to select a likely (arbitrary) value for capacitor C1 and algebraically rearrange the center frequency equation to solve for resistor R1:

$$F_c = \frac{1}{6.28 R_1 C_1}$$

$$R_1 = \frac{1}{6.28 F_c C_1}$$

Select a value of 0.047 µF (0.000 000 047 F) for capacitor C1. This means resistor R1 will need a value of

$$R_1 = \frac{1}{6.28 \times 880 \times 0.000\,000\,047}$$

$$= \frac{1}{0.000\,259\,7}$$

$$= 3\,850\ \Omega$$

You can round this off to the nearest standard resistor value, which happens to be 3.9 kΩ (3 900 Ω).

Resistor R2 has the same value as resistor R1 and capacitor

C2 is the same as capacitor C1:
$$R_2 = R_1$$
$$= 3.9 \text{ k}\Omega$$
$$C_2 = C_1$$
$$= 0.047 \text{ } \mu F$$

Next solve for resistor R3:
$$R_3 = \frac{R_1}{2}$$
$$= \frac{3\,900}{2}$$
$$= 1\,950 \text{ } \Omega$$

A 1.8-kΩ (1 800-Ω) resistor might be close enough, but in critical applications you might want to add a 150-Ω resistor in series with the 1.8-kΩ resistor.

Now find the value for capacitor C3:
$$C_3 = 2C_1$$
$$= 2 \times 0.047$$
$$= 0.094 \text{ } \mu F$$

You can round this off to 0.1 μF.

Then move on to the Q-determining components, starting with resistor R4:
$$R_4 = 2R_1Q$$
$$= 2 \times 3\,900 \times 4.4$$
$$= 34\,320 \text{ } \Omega$$

A standard 33-kΩ (33 000-Ω) resistor will probably be close enough. If the Q value is very critical, you can place a 1.2-kΩ (1 200-Ω) resistor in series with the 33-kΩ resistor.

Resistor R5 has the same value as resistor R4:
$$R_5 = R_4$$
$$= 33 \text{ k}\Omega$$

Finally, solve for capacitor C4:
$$C_4 = \frac{C_1}{Q}$$

$$= \frac{0.047}{4.4}$$

$$= 0.011 \, \mu F$$

You can round this off to a standard capacitor value of $0.01 \, \mu F$. Did this example maintain the necessary equalities?

$$Q = \frac{R_4}{2R_1} = \frac{C_1}{C_4}$$

Double-check the work:

$$\frac{R_4}{2R_1} = \frac{33\,000}{2 \times 3\,900}$$

$$= \frac{33\,000}{7\,800}$$

$$= 4.23$$

$$\frac{C_1}{C_4} = \frac{0.047}{0.01}$$

$$= 4.7$$

It is a little off due to the rounding off of the component values, but you would be reasonably close in both equations to the nominal Q value of 4.4.

The bridged differentiator

Another type of passive band-reject filter circuit is the bridged differentiator, shown in its most basic form in Fig. 6-5. The interesting feature about this filter circuit is that the reject band's center frequency can be manually adjusted via potentiometer R2. In effect, this potentiometer electrically looks like two resistors in series (R2a and R2b) with a ground connection made at their common junction.

The following component equalities must hold in this circuit:

- $R_1 = 6R_2$
- $C_1 = C_2 = C_3 = C$

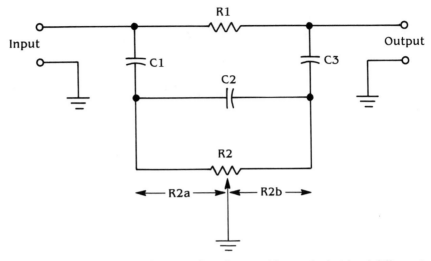

Fig. 6-5 *Another type of passive band-reject filter is the bridged differentiator.*

The center frequency of the reject band is equal to

$$F_c = \frac{1}{2\pi C \sqrt{3 R_{2a} R_{2b}}}$$

$$= \frac{1}{6.28 C \sqrt{3 R_{2a} R_{2b}}}$$

Notice that changing the position of the slider on potentiometer R2 changes the values of both R2a and R2b, and therefore the center frequency of the filter. This is best illustrated with an example.

Suppose you have a passive bridged differentiator band-reject filter circuit with the following component values:

- $R_1 = 150$ kΩ (150 000 Ω)
- $R_2 = 25$-kΩ potentiometer (25 000 Ω)
- $C_1, C_2, C_3 = 0.01$ μF (0.000 000 01 F)

Suppose the potentiometer's slider is set at its one-quarter point. This means that

$$R_{2a} = R_2 \times \frac{1}{4}$$

Active band-reject filters

$$R_{2b} = R_2 \times \frac{3}{4}$$

Working out the specific resistance values, we find

$$R_{2a} = 25\,000 \times \frac{1}{4}$$

$$= \frac{25\,000}{4}$$

$$= 6\,250\,\Omega$$

$$R_{2b} = 25\,000 \times \frac{3}{4}$$

$$= \frac{3 \times 25\,000}{4}$$

$$= \frac{75\,000}{4}$$

$$= 18\,750\,\Omega$$

This means the center frequency of the filter is

$$F_c = \frac{1}{6.28C\,\sqrt{3R_{2a}R_{2b}}}$$

$$= \frac{1}{6.28 \times 0.000\,000\,01 \times \sqrt{3 \times 6\,250 \times 18\,750}}$$

$$= \frac{1}{0.000\,000\,062\,8 \times \sqrt{351\,562\,500}}$$

$$= \frac{1}{0.000\,000\,062\,8 \times 18\,750}$$

$$= \frac{1}{0.001\,177\,5}$$

$$= 849\,\text{Hz}$$

Now look at what happens when you set the potentiometer's slider to the midpoint of its range. This time, the values of R_{2a} and R_{2b} become

$$R_{2a} = R_2 \times \frac{1}{2}$$

$$= 25\,000 \times \frac{1}{2}$$

$$= 12\,500\,\Omega$$

$$R_{2b} = R_2 \times \frac{1}{2}$$

$$= 25\,000 \times \frac{1}{2}$$

$$= 12\,500\,\Omega$$

This changes the center frequency of the filter to

$$F_c = \frac{1}{6.28 \times 0.000\,000\,01\sqrt{3 \times 12\,500 \times 12\,500}}$$

$$= \frac{1}{0.000\,000\,062\,8\sqrt{468\,750\,000}}$$

$$= \frac{1}{0.000\,000\,062\,8 \times 21\,650.635}$$

$$= \frac{1}{0.001\,359\,6}$$

$$= 735\,\text{Hz}$$

The bridged differentiator band-reject filter circuit can be adjusted for a wide range of center frequencies using just a single set of components. For this circuit to function properly, R_1 must be exactly equal to six times R_2. It is usually a good idea to use a trimpot in series with a fixed resistor for R_1 to precisely set the correct value.

In most practical bridged differentiator circuits, the center frequency control potentiometer will usually be flanked by a pair of fixed series resistors, as shown in the modified schematic diagram of Fig. 6-6. These changes do not really alter the design procedure in any significant way. All three capacitors must still have identical values and the series combination of R_1 and R_2 must equal six times the series combination of R_3, R_4, and R_5; that is,

$$R_1 + R_2 = 6(R_3 + R_4 + R_5)$$

Generally speaking, R_3 and R_5 will usually be given equal values, but this doesn't necessarily have to be so. These end resistors might be made unequal to force the tuneable range higher or lower.

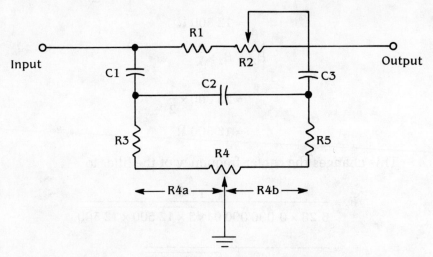

Fig. 6-6 *In most practical bridged differentiator circuits, the center frequency control potentiometer will usually be flanked by a pair of fixed series resistors.*

The center frequency equation can be rewritten as

$$F_c = \frac{1}{6.28C\sqrt{3R_aR_b}}$$

where

$$R_a = R_3 + R_{4a}$$

and

$$R_b = R_5 + R_{4b}$$

Otherwise there has been no functional change in the circuitry.

Figure 6-7 shows the circuit for an active bridged differentiator band-reject filter. Compare the input section of this circuit with the passive bridged differentiator circuits shown in Figs. 6-5 and 6-6. With the parts listed in Table 6-1, this filter circuit can be adjusted over a fairly wide range. The maximum center frequency is obtained when potentiometer R4's slider is set to the far left end of its range so $R_{3a} = 0\ \Omega$ and $R_{3b} = 25\ 000\ \Omega$. This means that

$$R_a = R_2 + R_{3a}$$
$$= 3\ 300 + 0$$
$$= 3\ 300\ \Omega$$

The bridged differentiator 147

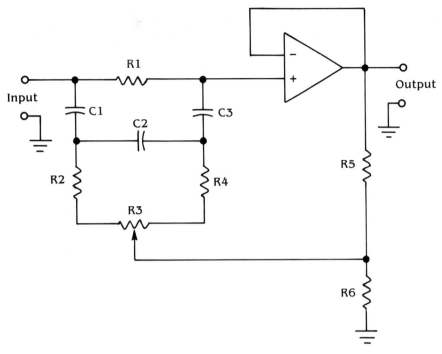

Fig. 6-7 *An active bridged differentiator band-reject filter circuit.*

Table 6-1 Suggested parts list for the active bridged differentiator band-reject filter circuit of Fig. 6-7.

R1	560 kΩ	(560 000 Ω)
R2	3.3 kΩ	(3 300 Ω)
R3	25-kΩ potentiometer	(25 000 Ω)
R4	68 kΩ	(68 000 Ω)
R5	68 Ω	
R6	2.2 kΩ	(2 200 Ω)
C1, C2, C3	0.05 μF	

and
$$R_b = R_4 + R_{3b}$$
$$= 68\,000 + 25\,000$$
$$= 93\,000 \ \Omega$$

The center frequency at this setting equals

$$F_c = \frac{1}{6.28C\sqrt{3R_aR_b}}$$

$$= \frac{1}{6.28 \times 0.000\,000\,05\sqrt{3 \times 3\,300 \times 93\,000}}$$

$$= \frac{1}{0.000\,000\,314\sqrt{920\,700\,000}}$$

$$= \frac{1}{0.000\,000\,314 \times 30\,343}$$

$$= \frac{1}{0.009\,527\,7}$$

$$= 105 \text{ Hz}$$

If you move the potentiometer's slider to its opposite extreme, you get the filter's minimum center frequency. Resistances R_{3a} becomes 25 000 Ω and R_{3b} becomes 0 Ω, so

$$R_a = 3\,300 + 25\,000$$
$$= 28\,300 \text{ Ω}$$

and

$$R_b = 68\,000 + 0$$
$$= 68\,000 \text{ Ω}$$

The center frequency now becomes

$$F_c = \frac{1}{6.28 \times 0.000\,000\,05\sqrt{3 \times 28\,300 \times 68\,000}}$$

$$= \frac{1}{0.000\,000\,314\sqrt{5\,733\,200\,000}}$$

$$= \frac{1}{0.000\,000\,314 \times 75\,981.577}$$

$$= \frac{1}{0.023\,858}$$

$$= 41.9 \text{ Hz}$$

The center frequency of this active bridged differentiator band-reject filter can be adjusted from about 42 Hz up to about 105 Hz, which is a pretty good range for a single manual control. The Q of this circuit is about 3.4, so the bandwidth is pretty narrow at such low center frequencies.

The active subtraction-type notch filter

Another way to create a band-reject filter circuit is to take a band-pass filter and subtract its output signal from the original (unfiltered) signal. A circuit that works like this is shown in Fig. 6-8. Notice that two op amp stages are required. The first op amp (along with resistors R1 through R3 and capacitors C1 and C2) is the band-pass filter. Because the inverting input of the op amp is used in this stage, the output from this op amp is inverted.

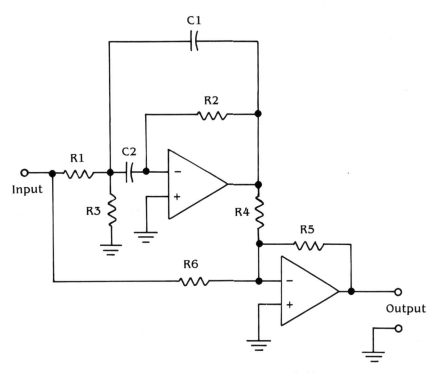

Fig. 6-8 An active subtraction-type notch filter circuit.

The second op amp (along with resistors R4 through R6) acts as a simple signal mixer combining the inverted output signal from the band-pass filter with the original, uninverted signal. All frequency components in the band-pass filter's passband are inverted and therefore their polarity is reversed. Positive becomes negative and vice versa. When these frequency components are mixed together with their uninverted counterparts, they cancel each other out and are omitted from the final output signal. On the other hand, those frequency components that are blocked by the band-pass filter stage are not cancelled out and they pass through the second op amp to the circuit's output.

The final output from this circuit is a mirror image of the output of the band-pass filter stage by itself. In other words, all filter components passed by the band-pass filter are blocked from the final output, and all frequency components blocked by the band-pass filter are included in the final output signal. We have a functional band-reject filter circuit.

This might seem like a silly, backwards way to go about it, but it does the job quite well with a minimum of component costs and design complications. This type of band-reject filter circuit might be simpler and more convenient to design than an equivalent band-reject filter circuit. The first section of this circuit (the band-pass filter) is really just the infinite gain multiple feedback band-pass filter discussed in some detail in chapter 5.

I review the design of this type of filter circuit. First, you must decide on the desired values for three operational values—the center frequency (F_c), the gain (K), and Q. In most practical design situations you will probably need to derive the Q value from the center frequency and the desired bandwidth (BW) of the filter's passband. Earlier, you saw that the formula for Q is

$$Q = \frac{F_c}{BW}$$

The first step in the actual design procedure is to arbitrarily select a likely value for C. (Remember $C_1 = C_2 = C$.) Then it's just a matter of solving for the three resistor values using the following formulas:

$$R_1 = \frac{Q}{2\pi F_c C K}$$

$$R_2 = \frac{2Q}{2\pi F_c C}$$

$$R_3 = \frac{Q}{2\pi F_c C (2Q - K)}$$

You can simplify your work a little by solving for an intermediate value I call Y:

$$Y = 2\pi F_c C$$

This subequation appears as part of each of the resistor equations, so you only have to calculate this part once, simplifying the resistor equations to

$$R_1 = \frac{Q}{YK}$$

$$R_2 = \frac{2Q}{Y}$$

$$R_3 = \frac{Q}{Y(2Q-K)}$$

The gain of the circuit (K) is controlled by the values of resistors R1 and R2. The formula is

$$K = \frac{R_2}{2R_1}$$

As a fairly typical example, assume that you want to design your filter circuit for a center frequency of 2 800 Hz (2.8 kHz), a bandwidth of 1 200 Hz (1.2 kHz), and the gain to unity (1). The first step is to determine the value of Q:

$$Q = \frac{F_c}{BW}$$

$$= \frac{2\ 800}{1\ 200}$$

$$= 2.3$$

Next, you need to select a likely value for C. In the audio range, 0.01 μF (0.000 000 01 F) is probably a pretty good choice. Now you need to solve for our intermediate variable (Y):

$$Y = 2\pi F_c C$$

$$= 6.28 F_c C$$

$$= 6.28 \times 2\ 800 \times 0.000\ 000\ 01$$

$$= 0.000\ 175\ 8$$

The remainder of the design procedure is just a matter of plugging your known values into the three resistor equations:

$$R_1 = \frac{Q}{YK}$$

$$= \frac{2.3}{0.000\,175\,8 \times 1}$$

$$= \frac{2.3}{0.000\,175\,8}$$

$$= 13\,083\,\Omega$$

$$R_2 = \frac{2Q}{Y}$$

$$= \frac{2 \times 2.3}{0.000\,175\,8}$$

$$= \frac{4.6}{0.000\,175\,8}$$

$$= 26\,166\,\Omega$$

$$R_3 = \frac{Q}{Y(2Q-K)}$$

$$= \frac{2.3}{0.000\,175\,8(2 \times 2.3 - 1)}$$

$$= \frac{2.3}{0.000\,175\,8(4.6 - 1)}$$

$$= \frac{2.3}{0.000\,175\,8 \times 3.6}$$

$$= \frac{2.3}{0.000\,632\,8}$$

$$= 3\,635\,\Omega$$

After rounding the resistances off to the nearest standard resistor values, the parts list for the example infinite gain multiple feedback band-pass filter circuit is as follows:

- $C_1 = 0.01\,\mu F$
- $C_2 = 0.01\,\mu F$
- $R_1 = 12\,k\Omega$ (12 000 Ω)
- $R_2 = 27\,k\Omega$ (27 000 Ω)
- $R_3 = 3.3\,k\Omega$ (3 300 Ω)

Better results will be achieved if you place a 330-Ω resistor in series with the 3.3-kΩ resistor for R3. Similarly, it might be a good idea to add a 1-kΩ resistor in series with the 12-kΩ resistor to give R1 a total value of 13 kΩ.

Well, this still leaves three resistors in the band-reject filter circuit to solve for—R_4, R_5, and R_7. To produce as deep a notch as possible, the following equality should be true:

$$\frac{R_6}{R_4} = \frac{R_2}{2R_1}$$

You can set a standardized value for either R_4 or R_6. Suppose you make R_4 equal to 10 kΩ (10 000 Ω). Now you can rearrange the equation to solve for R_6:

$$\frac{R_6}{10\,000} = \frac{R_2}{2R_1}$$

$$R_6 = \frac{10\,000 R_2}{2R_1}$$

In the example, you have already determined the values for resistors R1 and R2, so you can simply solve for R_6:

$$R_6 = \frac{10\,000 \times 27\,000}{2 \times 13\,000}$$

$$= \frac{270\,000\,000}{26\,000}$$

$$= 10\,385 \text{ Ω}$$

You can round off the value of R_6 to 10 kΩ (10 000 Ω), the same value as R_4. In most cases, these two resistors will turn out to have the same value, or close to it, especially if the band-pass filter section is designed for unity (1) gain.

The only component left in this circuit is the second op amp's feedback resistor (R5). The value of this resistor determines the gain of the passed frequencies at the output of the band-reject filter circuit. This op amp is wired as a simple inverting amplifier circuit (refer back to chapter 3 for more information). The standard gain formula for an inverting amplifier is:

$$K = \frac{-R_f}{R_i}$$

where
R_f = feedback resistor, and
R_i = input resistor.

In this case, the feedback resistor is R_5 and the input resistor is R_6, so

$$K = \frac{-R_5}{R_6}$$

The minus sign in this equation indicates that the signal polarity is inverted, which is irrelevent to the design and you can simply ignore it:

$$K = \frac{R_5}{R_6}$$

In a practical design situation, you will know the desired gain (K) and will need to solve for the resistance of R_5, so you can rearrange the equation:

$$R_5 = KR_6$$

In the example design, suppose you want a gain of about 5. You have already decided to use a 10-kΩ (10 000-Ω) resistor for R_6, so the required value for R_5 must be

$$R_5 = 5 \times 10\,000$$
$$= 50\,000\ \Omega$$

If you use a standard 47-kΩ (47 000-Ω) resistor for R_5, the actual gain will work out to

$$K = \frac{47\,000}{10\,000}$$
$$= 4.7$$

This should be close enough for most practical applications.

The VCVS active band-reject filter

Yet another common type of active filter circuit is shown in Fig. 6-9. This is the VCVS active band-reject filter. Other VCVS filter circuits were discussed in chapters 3, 4, and 5.

You might notice that this circuit resembles the active twin-T filter circuit of Fig. 6-4. You do have a twin-T in the input section of this circuit. One T (which happens to be shown upside down in the diagram) is composed of capacitors C1 and C2 and resistor R3. The second T is composed of resistors R1 and R2 along with capacitor C3.

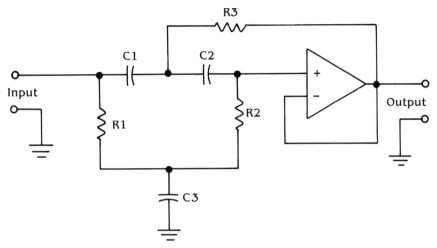

Fig. 6-9 *The band-reject filter function can also be achieved with a VCVS circuit.*

For this circuit to operate properly, capacitors C1 and C2 must have the same value, and capacitor C3 should have twice that value; that is

$$C_2 = C_1$$
$$C_3 = 2C_1$$

This type of filter circuit can be designed using the normalization method you have already worked with for many of the filter circuits in this book. It makes sense to standardize the normalized value for C_1 to 1 F. This will greatly simplify the design equations. This means you already know all of the normalized values for all of the capacitors in this circuit:

- $C_1 = C_2 = 1$ F
- $C_3 = 2C_1 = 2 \times 1 = 2$ F

In the design procedure, I will assume unity gain ($K = 1$).

Because this circuit has three resistors, you have three conductance values to solve for. All three of these conductances are derived from the desired Q value for the filter being designed:

$$G_1 = 2Q$$

$$G_2 = \frac{1}{2Q}$$

$$G_3 = 2Q + \frac{1}{2Q}$$

The next step in the design of this circuit is to find the frequency dependent normalization values:

$$u = 6.28F_c$$

$$ISF = \frac{F_c}{62.8}$$

The capacitor values are normalized in the usual way:

$$C = \frac{C_n}{(u)(ISF)}$$

Because C_1 and C_2 are normalized for a value of 1 F, the equation can be rewritten as

$$C_1 = C_2 = \frac{1}{(u)(ISF)}$$

Similarly, the standard normalization value for capacitor C3 is 2 F, so

$$C_3 = \frac{2}{(u)(ISF)}$$

Finding the actual resistor values for this circuit is simply a matter of multiplying the impedance scaling factor (ISF) by the reciprocal of the appropriate conductance value:

$$R_1 = \frac{ISF}{G_1}$$

$$= \frac{ISF}{2Q}$$

$$R_2 = \frac{ISF}{G_2}$$

$$= \frac{ISF}{\left(\frac{1}{2Q}\right)}$$

$$= ISF2Q$$

$$R_3 = \frac{ISF}{G_3}$$

$$= \frac{ISF}{\left(\dfrac{2Q+1}{2Q}\right)}$$

It is entirely possible that this design method will result in impractically low resistor values. If this should happen, just select a constant multiplier value (M). Usually 100 or 1 000 is a good choice for M, but any value can be used as long as it is used consistently throughout the circuit. Simply multiply each resistor value by the constant:

$$R_1 = MR_{1c}$$

$$R_2 = MR_{2c}$$

$$R_3 = MR_{3c}$$

where

R_{1c}, R_{2c}, and R_{3c} = original calculated resistance values.

At the same time, all of the capacitor values in the circuit must be divided by the same constant (M):

$$C_1 = \frac{C_{1c}}{M}$$

$$C_2 = \frac{C_{2c}}{M}$$

$$C_3 = \frac{C_{3c}}{M}$$

Try designing a typical VCVS band-reject filter circuit. In this example, you want your reject band to have a center frequency of 500 Hz and a bandwidth of 150 Hz. This means the desired Q of our example filter is

$$Q = \frac{F_c}{BW}$$

$$= \frac{500}{150}$$

$$= 3.333\,3$$

You already know the standard normalized values for the three capacitors in the circuit ($C_1 = C_2 = 1$ F and $C_3 = 2$ F), so you

Active band-reject filters

can go right to solving the conductance equations:

$$G_1 = 2Q$$
$$= 2 \times 3.333\,3$$
$$= 6.667 \text{ mhos}$$

$$G_2 = \frac{1}{2Q}$$
$$= \frac{1}{2 \times 3.333\,3}$$
$$= \frac{1}{6.667}$$
$$= 0.15 \text{ mho}$$

$$G_3 = 2Q + \frac{1}{2Q}$$
$$= (2 \times 3.333\,3) + \frac{1}{2 \times 3.333\,3}$$
$$= 6.667 + \frac{1}{6.667}$$
$$= 6.667 + 0.151\,5$$
$$= 6.817 \text{ mhos}$$

Next, you find your normalization factors:

$$u = 6.28 F_c$$
$$= 6.28 \times 500$$
$$= 3\,140$$

$$\text{ISF} = \frac{F_c}{62.8}$$
$$= \frac{500}{62.8}$$
$$= 7.96$$

Now you can denormalize the capacitor values:

$$C_1 = \frac{1}{(u)(\text{ISF})}$$
$$= \frac{1}{3\,140 \times 7.96}$$

$$= \frac{1}{25\,000}$$
$$= 0.000\,04 \text{ F}$$
$$= 40 \text{ }\mu\text{F}$$
$$C_2 = C_1$$
$$= 40 \text{ }\mu\text{F}$$
$$C_3 = \frac{2}{(u)(\text{ISF})}$$
$$= \frac{2}{3\,140 \times 7.96}$$
$$= \frac{2}{25\,000}$$
$$= 0.000\,08 \text{ F}$$
$$= 80 \text{ }\mu\text{F}$$

These capacitor values are rather high, but that's OK. The ISF and conductance values are also quite low, so you can expect the resistor values to turn out low too. Apply your multiplication constant (M), reducing the circuit capacitances and increasing the circuit resistances.

Solve for the resistor values and see if the prediction is accurate:

$$R_1 = \frac{\text{ISF}}{G_1}$$
$$= \frac{7.96}{6.667}$$
$$= 1.19 \text{ }\Omega$$

$$R_2 = \frac{\text{ISF}}{G_2}$$
$$= \frac{7.96}{0.15}$$
$$= 53.07 \text{ }\Omega$$

$$R_3 = \frac{\text{ISF}}{G_3}$$
$$= \frac{7.96}{6.817}$$
$$= 1.17 \text{ }\Omega$$

Yes, these resistor values are all far too low to be practical, so you will need to scale all of the component values. Select a multiplier (M) value of 2 000. This changes the component values in the circuit as follows;

$$R_1 = MR_{1c}$$
$$= 2\,000 \times 1.19$$
$$= 2\,380 \;\Omega$$
$$R_2 = MR_{2c}$$
$$= 2\,000 \times 53.07$$
$$= 106\,140 \;\Omega$$
$$R_3 = MR_{3c}$$
$$= 2\,000 \times 1.17$$
$$= 2\,340 \;\Omega$$
$$C_1 = \frac{C_{1c}}{M}$$
$$= \frac{40}{2\,000}$$
$$= 0.02 \;\mu F$$
$$C_2 = C_1$$
$$= 0.02 \;\mu F$$
$$C_3 = \frac{C_{3c}}{M}$$
$$= \frac{80}{2\,000}$$
$$= 0.04 \;\mu F$$

These component values are much more reasonable. In most cases, capacitor C3 can be made up of two capacitors of the same value as capacitor C1, connected in parallel. This way, you need four identical capacitors to build the circuit.

❖ 7
State variable and all-pass filters

IN THE PRECEDING CHAPTERS I HAVE COVERED THE FOUR BASIC types of filters—low-pass filters, high-pass filters, band-pass filters, and band-reject filters. These four types account for the vast majority of practical filter applications in modern electronics. But there are some specialized applications that call for other filter types. This chapter looks at two such "irregular" filter circuits—the state variable filter and the all-pass filter.

The basics of the state variable filter

A state variable filter is a sort of all-purpose filter circuit. It typically has three outputs, each performing a different filtering function on the same input signal. The following filter outputs are simultaneously available from a state variable filter:

- Low-pass
- High-pass
- Band-pass

In other words, the state variable filter does the job of three of the four basic filter types with a single circuit. For this reason, the state variable filter is sometimes called a universal filter.

A practical state variable filter circuit requires at least three op amp stages, one for each of the three simultaneous filter outputs. A block diagram of a state variable filter circuit is shown in

Fig. 7-1. Notice that this circuit has four basic elements:

- One summing block
- Two identical integrators
- One damping network

The high-pass output is taken off after the summing block, the band-pass output is taken off between the two integrators, and the low-pass output is the final output after the second integrator stage. The damping factor (usually identified as α) is equal to 1/Q of the band-pass filter section. This same damping factor is equally applied to all three outputs.

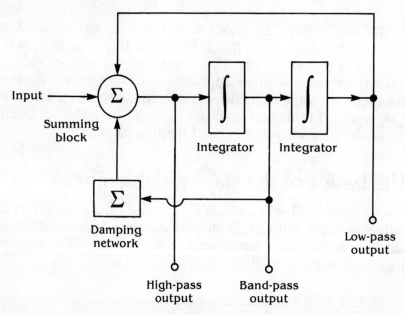

Fig. 7-1 *A state variable filter simultaneously performs multiple filter functions.*

The cutoff frequency at the low-pass output is identical to the cutoff frequency at the high-pass output. This same frequency is also the center frequency of the band-pass output. So you are dealing with one key frequency throughout the entire state variable filter. The other main variable in designing a state variable filter circuit is the damping factor (α), which controls the Q, and therefore the bandwidth, of the band-pass output and the roll-off slope for all three outputs.

The unity gain state variable filter

A practical state variable filter circuit is shown in Fig. 7-2. This is a unity gain state variable filter. Obviously, by definition, the gain of this circuit is one (unity). This type of filter circuit is not difficult to design because despite the relatively high parts count, only three passive component values are required. Both capacitors (C1 and C2) have the same value, which I call C. Eight of the nine resistors in the circuit (R1 through R8) have the same value, which I call R. Only resistor R9 has a different value.

The key frequency (cutoff frequency for the low-pass and high-pass sections and the center frequency for the band-pass section) is equal to

$$F_c = \frac{1}{2\pi RC}$$

$$= \frac{1}{6.28RC}$$

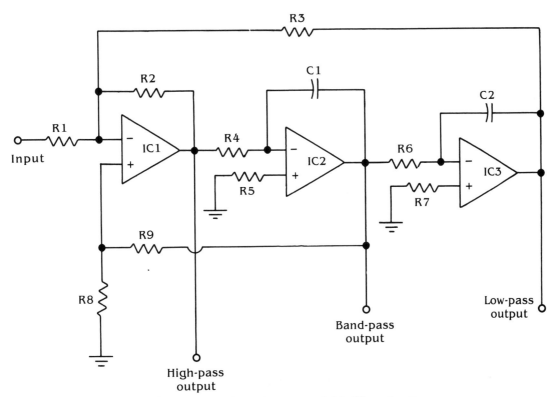

Fig. 7-2 *A unity gain state variable filter circuit.*

If the intended application is not too critical or if high-grade (low offset) op amps are used in the circuit, resistors R5 and R7 can sometimes be omitted. In this case, the noninverting inputs of IC2 and IC3 are shorted directly to ground. Personally, I feel it's worthwhile to include resistors R5 and R7 as cheap insurance against possible offset problems in the outputs. Resistors are very inexpensive and don't take up much space on a circuit board. In this particular circuit, no additional calculations are required to find the values for resistors R5 and R7 (they're the same as most of the other resistors in the circuit), so you might as well include them.

Resistors R8 and R9 make up the damping network of this state variable filter circuit. The value of resistor R9 is equal to

$$R_9 = \left(\frac{3}{\alpha} - 1\right)R_8$$

where

α = desired damping factor.

Earlier I stated that the damping factor (α) is always equal to the reciprocal of the filter Q; that is,

$$\alpha = \frac{1}{Q}$$

or

$$Q = \frac{1}{\alpha}$$

This allows you to rewrite the equation in the form

$$R_9 = (3Q - 1)R_8$$

In this way the circuit designer doesn't have to worry about the damping factor (α), but can work with the more familiar Q.

To design the unity gain state variable filter circuit, the designer first has to decide on the F_c and Q values. The next step is to arbitrarily select a likely value for capacitance (C) and rearrange the F_c equation to solve for R:

$$R = \frac{1}{6.28 F_c C}$$

Then, because $R_8 = R$, you can solve for resistor R9:

$$R_9 = (3Q - 1)R$$

Try a typical example. Select an F_c of 3 500 Hz (3.5 kHz) and a Q of 2. Choose a value of 0.01 µF (0.000 000 01 F) for C, and then solve for R:

$$R = \frac{1}{6.28 \times 3\,500 \times 0.000\,000\,01}$$

$$= \frac{1}{0.000\,219\,8}$$

$$= 4\,550\,\Omega$$

A 4.7-kΩ (4 700-Ω) resistor should be close enough for most practical purposes. Double-check and see how much this will throw your F_c value off from the desired 3 500 Hz:

$$F_c = \frac{1}{6.28RC}$$

$$= \frac{1}{6.28 \times 4\,700 \times 0.000\,000\,01}$$

$$= \frac{1}{0.000\,295\,1}$$

$$= 3\,389\text{ Hz}$$

Yes, that's pretty close. Component tolerances are likely to account for as much error as this rounding off of R causes.

Now all that's left to do is to find the required value for resistor R9:

$$R_9 = (3Q - 1)R$$
$$= (3 \times 2 - 1)\,4\,700$$
$$= (6 - 1)\,4\,700$$
$$= 5 \times 4\,700$$
$$= 23\,500\,\Omega$$

You can use a 22-kΩ (22 000-Ω) resistor, or for greater accuracy in the circuit's damping factor, a 22-kΩ resistor in series with a 1.5-kΩ (1 500-Ω) resistor for a total of 23 500 Ω.

There is one minor but important restriction on the use of this circuit. The Q must always be greater than 0.33. (This means the damping factor must be less than 3, because $\alpha = 1/Q$.) If the Q is too small, the value of R_9 will be negative, which is impossible. As an example of this, return to the example unity gain state variable filter circuit described above, but this time, assume that the desired Q is 0.25. This means the value of resistor R9 will be

$$R_9 = (3 \times 0.25 - 1)\,4\,700$$
$$= (0.75 - 1)\,4\,700$$
$$= -0.25 \times 4\,700$$
$$= -1\,175\,\Omega$$

Of course, a resistor with a negative value is impossible, so this cannot be built as a functional circuit. Fortunately, most practical filter applications call for Qs that are large enough to give R_9 a positive value, so this limitation will rarely be a serious problem.

An alternate approach to the design of this unity gain state variable filter circuit is the normalization method. Here, first normalize component values C_n and R_n. Capacitance C_n (C1 and C2) is normalized to 1 F. Resistance R_n (R_1 through R_8) is normalized to 1 Ω. Resistor R9 is not normalized. I'll get to this component in a few moments.

Because this circuit has no other component calculations, you can proceed directly to finding the frequency normalization factor (u) and the impedance scaling factor (ISF):

$$u = 2\pi F_c$$
$$= 6.28 F_c$$

$$\text{ISF} = \frac{F_c}{20\pi}$$
$$= \frac{F_c}{62.8}$$

The formula for denormalizing the capacitance value is

$$C = \frac{C_n}{(u)(\text{ISF})}$$

Because you know that C_n has been normalized to 1 F, this equation becomes

$$C = \frac{1}{(u)(\text{ISF})}$$

Because both capacitors (C1 and C2) in this circuit have the same value, you only have to perform this calculation once.

To denormalize the resistance values, use the formula:

$$R = (R_n)(\text{ISF})$$

Of course, you already know that R_n is normalized to 1 Ω, so you can simplify this equation to

$$R = 1 \times \text{ISF}$$
$$= \text{ISF}$$

The resistance value (R) is equal to the impedance scaling factor (ISF) in this circuit. Because this circuit has only one normalized resistance, you don't have to perform any more calculations.

In some cases, the value of R will come out too low to be practical. If this should be the case in your particular design, simply multiply R by a constant (M) (usually but not necessarily 100 or 1 000) and divide C by the same constant:

$$R = RM$$

$$C = \frac{C}{M}$$

Finally, you need to solve for the value of resistor R9. This is done using the exact same formula used in the earlier design method:

$$R_9 = (3Q - 1)R$$

As you can see, while this circuit looks fairly complex, with a rather high parts count, it is actually relatively simple to design.

Try an example. This time design a unity gain state variable filter circuit with an F_c of 5 250 Hz (5.25 kHz) and a Q of 8.4. Because the values of R_n and C_n are always normalized to one, jump straight to finding the normalization factors for your desired F_c:

$$u = 6.28 F_c$$
$$= 6.28 \times 5\,250$$
$$= 32\,970$$

$$\text{ISF} = \frac{F_c}{62.8}$$
$$= \frac{5\,250}{62.8}$$
$$= 84$$

Now you can denormalize the capacitor value:

$$C = \frac{1}{(u)(\text{ISF})}$$

$$= \frac{1}{32\,970 \times 84}$$

$$= \frac{1}{2\,769\,480}$$

$$= 0.000\,000\,36 \text{ F}$$

$$= 0.36 \ \mu\text{F}$$

The denormalized resistance value (R) is equal to the impedance scaling factor (ISF):

$$R = \text{ISF}$$
$$= 84 \ \Omega$$

This is rather low, so use a multiplication constant. This time, make M equal to 150:

$$R = 84 \times 150$$
$$= 12\,600 \ \Omega$$

$$C = \frac{0.36}{150}$$

$$= 0.002\,4 \ \mu\text{F}$$

You can use 12-kΩ (12 000-Ω) resistors for R1 through R8 and 0.002 2-μF capacitors for C1 and C2.

Now all you have to do is find the required value for resistor R9:

$$R_9 = (3Q - 1)R$$
$$= (3 \times 8.4 - 1)\,12\,000$$
$$= (25.2 - 1)\,12\,000$$
$$= 24.2 \times 12\,000$$
$$= 290\,400 \ \Omega$$

If the application isn't too critical, you can use a standard 270-kΩ (270 000 Ω) resistor for R_9. For greater accuracy, R_9 could be a series combination of a 270-kΩ resistor and a 22-kΩ (22 000-Ω) resistor, for a total of 292 kΩ (292 000 Ω).

The four op amp state variable filter

An improved state variable filter circuit is shown in Fig. 7-3. Compare this circuit to the one shown in Fig. 7-2. Notice that I have added a fourth op amp stage. This fourth op amp stage is

The four op amp state variable filter 169

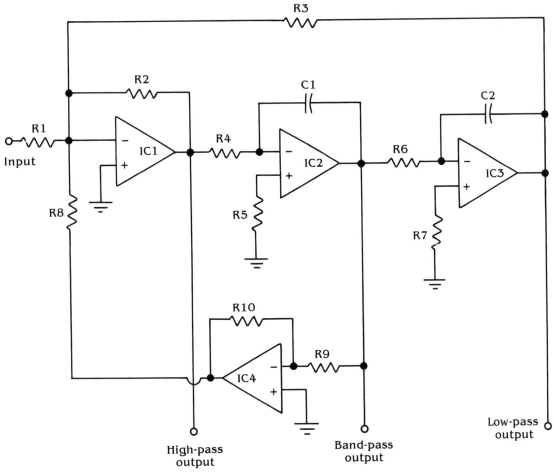

Fig. 7-3 *Improved performance can be achieved with the four op amp state variable filter circuit.*

basically just an inverting amplifier. It permits you to design your state variable filter for greater than unity gain. Remember, the circuit of Fig. 7-2 is restricted to unity gain. Moreover, the gain and the damping factor in the circuit of Fig. 7-3 are independent of one another.

The main body of this circuit is quite similar to the unity gain state variable filter circuit discussed over the last few pages. The same F_c formula is used for this circuit:

$$F_c = \frac{1}{2\pi RC}$$

$$= \frac{1}{6.28RC}$$

Of course, F_c is the cutoff frequency for both the low-pass and high-pass outputs, as well as the center frequency for the band-pass output. All three filter outputs must always use the same F_c value. Resistors R2 through R9 all have the same value which we will call R. Both capacitor C1 and C2 have a value of C. I get to resistors R1 and R10 shortly.

In a practical circuit design situation, begin by selecting a likely capacitance value for C and then rearrange the frequency equation to solve for R:

$$R = \frac{1}{6.28 F_c C}$$

Resistor R1 is selected to set the gain of the circuit according to the formula

$$R_1 = \frac{R}{K}$$

where
  R = value of resistors R2 through R9, and
  K = desired gain.

The new, added op amp stage (IC4, R9, and R10) determines the damping factor and thus the Q of the filter. Resistor R9's value, of course, is simply R. The formula for R_{10} is:

$$R_{10} = \alpha R$$

where
  α = damping factor.

Because the damping factor is the reciprocal of Q,

$$\alpha = \frac{1}{Q}$$

you can rewrite the R_{10} equation as

$$R_{10} = \left(\frac{1}{Q}\right) R$$

$$= \frac{R}{Q}$$

As you can see, the design equations for this circuit are quite straightforward and not difficult to use or understand. Now work through a typical design example to become more familiar with these formulas. You will be designing a four op amp state variable filter circuit with the following characteristics:

- F_c = 6 900 Hz (6.9 kHz)
- BW = 800 Hz (0.8 kHz)
- K = 5

The Q of this circuit is determined by F_c and the bandwidth as with any filter circuit:

$$Q = \frac{F_c}{\text{BW}}$$

$$= \frac{6\,900}{800}$$

$$= 8.625$$

The damping factor (α) is equal to the reciprocal of the circuit's Q:

$$\alpha = \frac{1}{Q}$$

$$= \frac{1}{8.625}$$

$$= 0.115\,9$$

Select a 0.002 2-μF (0.000 000 002 2-F) capacitor as the C value to begin the actual design. This choice, remember, is arbitrary. If after working the equation, you find you have an unsuitable value for R, simply select a new value for C and try the equation again. Using a 0.002 2-μF capacitor for C gives the following value for R:

$$R = \frac{1}{6.28 F_c C}$$

$$= \frac{1}{6.28 \times 6\,900 \times 0.000\,000\,002\,2}$$

$$= \frac{1}{0.000\,086\,66}$$

$$= 1\,539\ \Omega$$

You can use a common 10-kΩ (10 000-Ω) resistors for R2 through R9.

Use your desired gain value ($K = 5$) to find the value of resistor R1:

$$R_1 = \frac{R}{K}$$

$$= \frac{10\,000}{5}$$

$$= 2\,000 \text{ Ω}$$

You can probably use a 2.2-kΩ (2 200-Ω) resistor for R1, although this would throw the circuit gain off slightly:

$$R_1 = \frac{R}{K}$$

so

$$K = \frac{R}{R_1}$$

$$= \frac{10\,000}{2\,200}$$

$$= 4.545\,4$$

If this difference is critical for your particular application, you can use two low-tolerance 1-kΩ (1 000-Ω) resistors in series for R1, giving a nominal total value of 2 000 Ω.

Finally, solve for the value of resistor R10 based on the desired Q for the finished circuit:

$$R_{10} = \frac{R}{Q}$$

$$= \frac{10\,000}{8.625}$$

$$= 1\,159 \text{ Ω}$$

You could use a 1-kΩ (1 000-Ω) resistor or a 1.2-kΩ (1 200-Ω) resistor for R10. Alternatively, you could use a 1-kΩ (1 000-Ω) resistor in series with a 150-Ω resistor.

You can also design this type of circuit using the normalization method. In this method, normalize the value of R_n to 1 Ω and C_n to 1 F. Don't worry about resistors R1 and R10 until you have denormalized R.

Begin this design by finding the normalization factors:

$$u = 2\pi F_c$$
$$= 6.28 F_c$$

$$\text{ISF} = \frac{F_c}{20\pi}$$

$$= \frac{F_c}{62.8}$$

The usual formula is used to denormalize the capacitance value:

$$C = \frac{C_n}{(u)(\text{ISF})}$$

$$= \frac{1}{(u)(\text{ISF})}$$

The denormalized resistance value (R) has a value of

$$R = R_n \times \text{ISF}$$
$$= 1 \times \text{ISF}$$
$$= \text{ISF}$$

The denormalized resistance (R) is equal to the impedance scaling factor (ISF).

This method is likely to result in an inconveniently low R, so multiply R by a constant (M) and divide C by the same constant:

$$R = RM$$

$$C = \frac{C}{M}$$

Solve for resistors R1 and R10 using the same formulas from the earlier design method:

$$R_1 = \frac{R}{K}$$

$$R_{10} = \frac{R}{Q}$$

Work through a typical example and design a four op amp state variable filter circuit with the following characteristics:

- $F_c = 4\,200$ Hz (4.2 kHz)
- BW = $1\,700$ Hz (1.7 kHz)
- $K = 6.6$

The Q of this circuit is determined by F_c and the bandwidth as with any filter circuit:

$$Q = \frac{F_c}{BW}$$

$$= \frac{4\,200}{1\,700}$$

$$= 2.47$$

For convenience round this off to 2.5, which shouldn't noticeably affect the performance of the finished circuit to any significant degree.

The damping factor (α) is equal to the reciprocal of the circuit's Q:

$$\alpha = \frac{1}{Q}$$

$$= \frac{1}{2.5}$$

$$= 0.4$$

Always normalize R_n to 1 Ω, and C_n to 1 F, so you can jump right in and find the frequency normalization factor (u):

$$u = 6.28F_c$$
$$= 6.28 \times 4\,200$$
$$= 26\,376$$

Next, calculate the impedance scaling factor (ISF):

$$ISF = \frac{F_c}{62.8}$$

$$= \frac{4\,200}{62.8}$$

$$= 66.879$$

You can round off the ISF value to 67. Now you can denormalize the capacitance value:

$$C = \frac{1}{(u)(ISF)}$$

$$= \frac{1}{26\,376 \times 67}$$

$$= \frac{1}{1\,767\,192}$$

$$= 0.000\,000\,57 \text{ F}$$

$$= 0.57 \text{ } \mu\text{F}$$

The nominal denormalized value of R is equal to the impedance scaling factor:

$$R = \text{ISF}$$
$$= 67 \text{ }\Omega$$

But this is too low to be really practical, so multiply the resistance by 1 010 and divide the capacitance by the same 1 010. (This unusual multiplier factor was selected by trial and error to result in close to standard component values for both R and C.)

$$R = RM$$
$$= 67 \times 1\,010$$
$$= 67\,670 \text{ }\Omega$$

$$C = \frac{C}{M}$$

$$= \frac{0.57}{1\,010}$$

$$= 0.000\,56 \text{ } \mu\text{F}$$

$$= 560 \text{ pF}$$

You can use 68-kΩ (68 000-Ω) resistors for R2 through R9 and 560-pF capacitors for C1 and C2. If your application isn't too critical about the exact value of F_c, you might be able to substitute 500-pF (or 470-pF) capacitors, which are likely to be easier to find.

Next, find the value of the gain-determining resistor (R1):

$$R_1 = \frac{R}{K}$$

$$= \frac{68\,000}{6.6}$$

$$= 10\,303 \text{ }\Omega$$

A standard 10-kΩ (10 000-Ω) resistor should do fine for R_1.

The last step in the design is to find the required value for

R_{10}, which is based on the damping factor or the Q of the desired filter:

$$R_{10} = \frac{R}{Q}$$
$$= \frac{68\,000}{2.5}$$
$$= 27\,200 \; \Omega$$

This is very close to 27 kΩ (27 000 Ω), which is a standard resistor value.

State variable band-reject filters

Basic state variable filters offer low-pass, high-pass, and band-pass outputs, but generally they do not support the band-reject filter function. However, a state variable band-reject filter is possible with just a little added circuitry.

In chapter 2, you saw that a band-reject filter could be created by placing a low-pass filter and a high-pass filter in parallel. Similarly, in a state variable band-reject filter circuit, the band-reject output is derived from the low-pass and high-pass outputs, albeit in a different way. Basically, in a state variable band-reject filter, the low-pass and high-pass outputs are summed together at the input of an added inverting amplifier stage, as illustrated in the block diagram of Fig. 7-4.

For this to work, the low-pass and high-pass signals must be subjected to exactly equal gains. Usually, all three of the resistors in this inverting amplifier stage are given equal values for unity

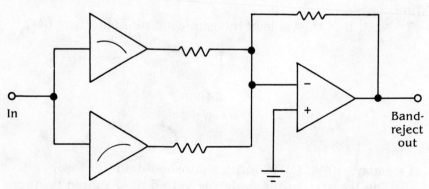

Fig. 7-4 Some state variable filter circuits also have a band-reject output.

gain, but as long as the two input resistors are equal, the feedback resistor can be changed for different gains. By using unity gain, the depth of the notch at the center frequency (F_c) will be about − 30 dB, with respect to the passbands on either side of the reject band.

The same basic approach can be used with discrete low-pass and high-pass filter circuits. Their outputs can be summed together to create a band-reject filter output. However, both filter circuits must be very closely matched or the resulting reject band will not be symmetrical around the nominal center frequency (F_c). Of course, when a state variable filter circuit is used, the low-pass and high-pass sections will always be identically matched because they are both part of the same circuit.

Figure 7-5 shows the complete schematic diagram for a typical state variable band-reject filter circuit. Depending on the intended application, the high-pass, low-pass, or band-pass outputs may also be tapped off from this circuit. If either the low-pass or the high-pass output (or both) is used simultaneously with the band-reject output, there might be some problems with loading down the circuit. Buffer amplifiers between the filter outputs and the circuits they are driving might be advisable or even necessary.

The state variable band-reject filter circuit is designed in just the same way as the basic state variable filter circuit discussed earlier in this chapter. The only differences are the additions of IC4, R10, R11, and R12. Resistors R10 and R11 are the input resistors to the summing amplifier stage that produces the band-reject output. These two resistors must have the same value. It doesn't matter too much what resistance value is used here, as long as the low-pass signal (through R10) and the high-pass signal (through R11) are fed through equal resistances. A pair of 10-kΩ (10 000-Ω) resistors are good general-purpose choices for R10 and R11.

The value of feedback resistor R12 determines the gain of this stage. The gain formula for an inverting amplifier circuit is

$$K = \frac{-R_{12}}{R_{10}}$$

where it is assumed that

$$R_{10} = R_{11}$$

The minus sign in the gain equation indicates the polarity inver-

178 State variable and all-pass filters

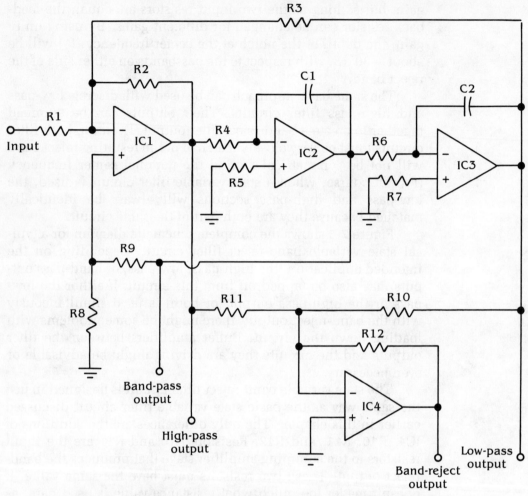

Fig. 7-5 *A typical state variable filter circuit with a band-reject output.*

sion of any signal fed through an inverting amplifier. For these purposes, you can ignore this polarity inversion. If you standardize the value of R_{10} (and R_{11}) to 10 kΩ (10 000 Ω), then you can rewrite the equation to solve for the value of R_{12} from the desired gain value

$$R_{12} = R_{10}K$$
$$= 10\ 000K$$

In most practical applications for a state variable band-reject filter circuit like this, you will probably want the band-reject output to have the same gain as the rest of the filter. Therefore, this stage should have unity gain (K = 1). This means that the value of resistor R12 should be

$$R_{12} = 10\,000 \times 1$$
$$= 10\,000\ \Omega$$

In most cases, R12 (like R10 and R11) should be a 10-kΩ (10 000-Ω) resistor. You can standardize these values and not worry about coming up with new values when designing specific circuits according to this pattern.

The rest of this circuit is simply the standard state variable filter circuit described earlier in this chapter. It is designed in the same way as before using either the F_c equation or the normalization method.

I go through two quick examples using each of these design methods. First, design a state variable band-reject filter with an F_c of 7 500 Hz (7.5 kHz) and a bandwidth of 3 000 Hz (3 kHz). This is, by definition, a unity gain state variable filter, so K = 1.

As you should recall, this type of filter circuit is not difficult to design because, despite the relatively high parts count, only three passive component values are required. Both capacitors (C1 and C2) have the same value, which I call C. Eight of the nine remaining resistors in the circuit (R1 through R8) have the same value, which I call R. Only resistor R9 has a different value.

The key frequency (cutoff frequency for the low-pass and high-pass sections and the center frequency for the band-pass section) is equal to

$$F_c = \frac{1}{2\pi RC}$$
$$= \frac{1}{6.28RC}$$

In designing the example circuit, select 0.004 7 μF (0.000 000 004 7 F) as our value for C and rearrange the F_c equation to solve for R:

$$R = \frac{1}{6.28 F_c C}$$
$$= \frac{1}{6.28 \times 7\,500 \times 0.000\,000\,004\,7}$$
$$= \frac{1}{0.000\,188\,4}$$
$$= 5\,308\ \Omega$$

You can use standard 4.7-kΩ (4 700-Ω) resistors for R1 through R8.

Resistors R8 and R9 make up the damping network of the state variable filter circuit. The value of resistor R9 is equal to

$$R_9 = \left(\frac{3}{\alpha} - 1\right)R_8$$

where
α = desired damping factor.

Earlier I stated that the damping factor (α) is always equal to the reciprocal of the filter Q; that is,

$$\alpha = \frac{1}{Q}$$

or

$$Q = \frac{1}{\alpha}$$

This allows you to rewrite the equation in the form

$$R_9 = (3Q - 1)R_8$$

The circuit designer doesn't have to worry about the damping factor (α), but can work with the more familiar Q, which works out to

$$Q = \frac{F_c}{BW}$$

$$= \frac{7\,500}{3\,000}$$

$$= 2.5$$

You can complete the design by finding the correct value for resistor R9:

$$\begin{aligned} R_9 &= (3 \times 2.5 - 1) \times 4\,700 \\ &= (7.5 - 1) \times 4\,700 \\ &= 6.5 \times 4\,700 \\ &= 30\,550 \ \Omega \end{aligned}$$

You can use either a 27-kΩ (27 000-Ω) resistor or a 33-kΩ (33 000-Ω) resistor for R_9. To come even closer to the nominal value, you could try a 27-kΩ resistor in series with a 3.3-kΩ resistor for a total of 30 300 Ω, which is quite close to the calculated value.

The completed example state variable band-reject filter circuit calls for the following parts:

- $R_1 - R_8 = 4.7$ kΩ (4 700 Ω)
- $R_9 = 30.3$ kΩ (30 300 Ω)
    $= 27$ k$\Omega + 3.3$ kΩ in series
- $R_{10} - R_{12} = 10$ kΩ (10 000 Ω)
- $C_1, C_2 = 0.004\ 7$ μF

Now try a second example, this time using the normalization method. This time design the state variable band-reject filter circuit for a center frequency (F_c) of 11 500 Hz (11.5 kHz) and a bandwidth of 2 200 Hz (2.2 kHz). This means that the filter's Q is equal to

$$Q = \frac{F_c}{BW}$$

$$= \frac{11\ 500}{2\ 200}$$

$$= 5.2$$

Once again, resistors R10, R11, and R12 all have standard values of 10 kΩ (10 000 Ω), so you don't need to concern yourselves with these components in the design procedure. Resistors R1 through R8 all have the same value, which I identify as R_n. The standard normalization value for R_n is 1 Ω. Similarly, capacitors C1 and C2 have identical values (C_n), which we give a standardized normalization value of 1 F. Resistor R9 is not normalized. We'll get to this component in a few moments.

Because this circuit has no other component calculations, you can proceed directly to finding the frequency normalization factor (u) and the impedance scaling factor (ISF):

$$u = 2\pi F_c$$
$$= 6.28 F_c$$
$$= 6.28 \times 11\ 500$$
$$= 72\ 220$$

$$ISF = \frac{F_c}{20}$$

$$= \frac{F_c}{62.8}$$

$$= \frac{11\,500}{62.8}$$

$$= 183$$

The formula for denormalizing the capacitance value is

$$C = \frac{C_n}{(u)(\text{ISF})}$$

Because you know that C_n has been normalized to 1 F, the equation becomes

$$C = \frac{1}{(u)(\text{ISF})}$$

$$= \frac{1}{72\,220 \times 183}$$

$$= \frac{1}{13\,216\,260}$$

$$= 0.000\,000\,076$$

$$= 0.076$$

Because both capacitors (C1 and C2) in this circuit have the same value, you only have to perform this calculation once.

To denormalize the resistance values, use the formula

$$R = (R_n)(\text{ISF})$$

Of course, you already know that R_n is normalized to 1 Ω, so you can simplify this equation to

$$R = 1 \times \text{ISF}$$
$$= \text{ISF}$$
$$= 183\ \Omega$$

The resistance value (R) is equal to the impedance scaling factor (ISF) in this circuit. Because this circuit only has one normalized resistance, you don't have to perform any more calculations.

As is often the case when using the normalization design method, the value of R has come out a bit too low to be practical. When this happens, all you have to do is multiply R by a constant (M) (usually but not necessarily 100 or 1 000), and divide C by the same constant:

$$R = RM$$

$$C = \frac{C}{M}$$

In this case, I use a multiplier value of 150, so

$$R = 183 \times 150$$
$$= 27\,450\ \Omega$$

$$C = \frac{0.076}{150}$$
$$= 0.000\,5\ \mu F$$
$$= 500\ pF$$

You can use standard 27-kΩ (27 000-Ω) resistors for R (R1 through R8) and 500-pF (or 470-pF) capacitors for C (C1 and C2).

Now all you have to do to complete the design is to determine the value of resistor R9. This is done in exactly the same way as in the other design method for this circuit:

$$R_9 = (3Q - 1)R$$
$$= (3 \times 5.2 - 1) \times 27\,000$$
$$= (15.6 - 1) \times 27\,000$$
$$= 14.6 \times 27\,000$$
$$= 394\,200\ \Omega$$

This is very close to 390 kΩ (390 000 Ω), which is a standard resistor value.

The finished parts list for the second state variable band-reject filter circuit design example is as follows:

- $R_1 - R_8 = 27$ kΩ (27 000 Ω)
- $R_9 = 390$ kΩ (390 000 Ω)
- $R_{10} - R_{12} = 10$ kΩ (10 000 Ω)
- $C_1, C_2 = 500$ pF (or 470 pF)

Adding the band-reject function to the basic state variable filter circuit does not noticeably increase the complexity of the circuit design procedure because the added components (IC4, R10, R11, and R12) are given standardized values, regardless of the filter's desired F_c or Q values.

The same approach can be used to add the band-reject function to a four op amp state variable filter circuit. Of course, you now have a five op amp circuit. The revised schematic for this circuit is shown in Fig. 7-6. The main body of the filter circuit (including capacitors C1 and C2 and resistors R1 through R8) is designed in the same way as in the ordinary state variable filter circuit described earlier in this chapter. The band-reject function

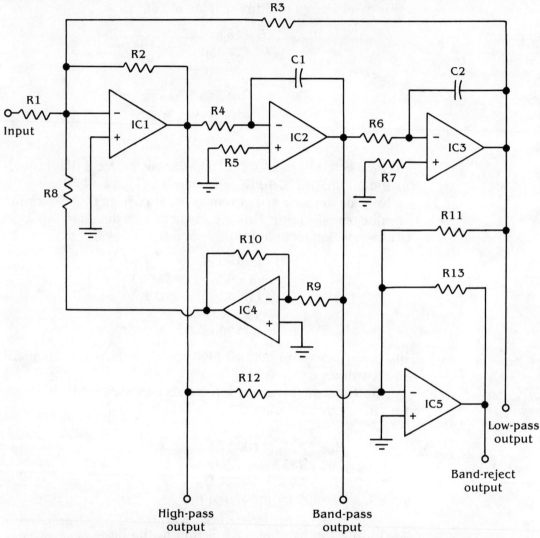

Fig. 7-6 *This four op amp state variable filter circuit includes the band-reject function.*

is added with an inverting summing amplifier stage made up of IC5 and resistors R11, R12, and R13.

Resistors R11 and R12 are the input resistors to the summing amplifier stage that produces the band-reject output. These two resistors must have the same value. It doesn't matter what resistance value is used here, as long as the low-pass signal (through R11) and the high-pass signal (through R12) are fed through equal resistances. A pair of 10-kΩ (10 000-Ω) resistors are good general-purpose choices for R11 and R12.

The value of feedback resistor R13 determines the gain of this stage. The gain formula for an inverting amplifier circuit is

$$K = \frac{-R_{13}}{R_{11}}$$

where it is assumed that

$$R_{12} = R_{11}$$

The minus sign in the gain equation indicates the polarity inversion of any signal fed through an inverting amplifier. For these purposes, you can ignore this polarity inversion. If you standardize the value of R_{11} (and R_{12}) to 10 kΩ (10 000 Ω), then you can rewrite the equation to solve for the value of R_{13} from the desired gain value:

$$R_{13} = R_{11}K$$
$$= 10\,000K$$

In most practical applications for a state variable band-reject filter circuit like this, you would probably want the band-reject output to have the same gain as the rest of the filter. Therefore, this stage should have unity gain ($K = 1$). This means that the value of resistor R13 should be

$$R_{13} = 10\,000 \times 1$$
$$= 10\,000\,\Omega$$

In most cases, R_{13} (like R_{11} and R_{12}) should be a 10-kΩ (10 000-Ω) resistor. You can standardize these values and not worry about coming up with new values when designing specific circuits according to this system.

The all-pass filter

One of the most unusual types of filter circuits is the all-pass filter. In fact, you could even say that this one is downright peculiar. As the name indicates, this type of "filter" passes everything. All of the input signal's frequency components get through—nothing is blocked. It doesn't sound like much of a filter at all. So what good is it?

While all of the frequency components passing through an all-pass filter circuit are subjected to equal attenuation (usually designed to be minimal), they are also subjected to equal phase shift. In other words, an all-pass filter is used where a controlled varying phase response is required. Another common name for this type of circuit is the phase-shift filter. That is probably a better name, because it gives a clearer description of the circuit's function.

All-pass filters are often used in equalizers and compensation networks, among other specialized applications. Figure 7-7 shows one of the simplest all-pass filter circuits. This is a multiple feedback all-pass filter.

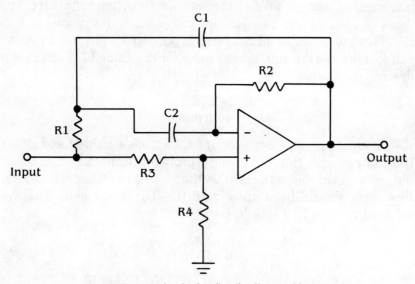

Fig. 7-7 A multiple feedback all-pass filter circuit.

Two intermediate variables (a and b) are required in the design procedure of a multiple feedback all-pass filter circuit. Unfortunately, the required mathematics for this design are a bit more complex than most of the other filter circuits you have been dealing with.

In an all-pass filter, you are not interested in a cutoff frequency (F_c) but in a specific amount of phase shift (w). The phase shift must be specified for a specific frequency, which I might as well call F_c. All frequency components passing through the all-pass filter will be shifted by an equal amount. This is just a reference frequency that will be used later for denormalizing the

Table 7-1 Common tangent values.

x	Tan(x)	x	Tan(x)
0	0.0000	46	1.0355
1	0.0175	47	1.0724
2	0.0349	48	1.1106
3	0.0524	49	1.1504
4	0.0699	50	1.1918
5	0.0875	51	1.2349
6	0.1051	52	1.2799
7	0.1228	53	1.3270
8	0.1405	54	1.3764
9	0.1584	55	1.4281
10	0.1763	56	1.4826
11	0.1944	57	1.5399
12	0.2126	58	1.6003
13	0.2309	59	1.6643
14	0.2493	60	1.7321
15	0.2679	61	1.8040
16	0.2867	62	1.8807
17	0.3057	63	1.9626
18	0.3249	64	2.0503
19	0.3443	65	2.1445
20	0.3640	66	2.2460
21	0.3839	67	2.3559
22	0.4040	68	2.4751
23	0.4245	69	2.6051
24	0.4452	70	2.7475
25	0.4663	71	2.9042
26	0.4877	72	3.0777
27	0.5095	73	3.2709
28	0.5317	74	3.4874
29	0.5543	75	3.7321
30	0.5774	76	4.0108
31	0.6009	77	4.3315
32	0.6249	78	4.7046
33	0.6494	79	5.1446
34	0.6745	80	5.6713
35	0.7002	81	6.3138
36	0.7265	82	7.1154
37	0.7536	83	8.1443
38	0.7813	84	9.5144
39	0.8098	85	11.43
40	0.8391	86	14.30
41	0.8693	87	19.08
42	0.9004	88	28.64
43	0.9325	89	57.29
44	0.9657	90	∞ (infinity)
45	1.0000		

component values in the circuit. The phase shift (at F_c) is equal to

$$w = -2\tan^{-1}\left(\frac{a}{(b-1)}\right)$$

where
   $\tan^{-1}$ = antitangent.

Common tangent values are listed in Table 7-1. As you shall soon see, you won't need to actually bother with any antitangents.

With a little rearranging, the equation looks like

$$\tan\left(\frac{w}{-2}\right) = \frac{a}{(b-1)}$$

Therefore,

$$a = (b-1) \times \tan\left(\frac{w}{-2}\right)$$

or

$$b = 1 + \frac{a}{\tan\left(\frac{w}{-2}\right)}$$

To simplify the appearance of the later equations, I assign a new intermediate variable, which I call Y:

$$Y = \tan\left(\frac{w}{-2}\right)$$

By substituting Y into the above equations, you get

$$a = (b-1)Y$$
$$b = 1 + a(Y)$$

The gain of the circuit (K) is also defined by variables a and b:

$$K = \frac{b}{(a^2 + b)}$$

Again, you can rearrange the formula as follows:

$$\frac{b}{K} = a^2 + b$$

$$\frac{b}{K} - b = a^2$$

$$a^2 = \frac{b}{K} - b$$

$$a^2 = \left(\frac{1}{K} - 1\right)b$$

You already derived an earlier equation to define b in terms of a:

$$b = 1 + a(Y)$$

You can combine and rearrange the two equations to solve for a:

$$a^2 = \left(\frac{1}{K-1}\right)(1 + a(Y))$$

At this point, you have a on both sides of the equals sign, but you really can't go any further until you find specific values for Y and K. Once you know these two variables, you can plug them into the above equation and solve for a. Then, knowing a, you can solve for b:

$$b = 1 + a(Y)$$

The next step is to normalize the value of C to 1 F. Notice that the same capacitance value (C) is used for both C_1 and C_2 in this circuit. You can then use the values of a, b, and C to find the normalized values for the four resistors in the circuit:

$$R_{1n} = \frac{a}{2bC}$$

$$R_{2n} = \frac{2}{aC}$$

$$R_{3n} = \frac{2(a^2 + b)}{abC}$$

$$R_{4n} = \frac{2(a^2 + b)}{a^2 C}$$

But because C is normalized to a value of 1 F, it can be eliminated from each of these equations, leaving

$$R_{1n} = \frac{a}{2b}$$

$$R_{2n} = \frac{2}{a}$$

$$R_{3n} = \frac{2(a^2 + b)}{ab}$$

$$R_{4n} = \frac{2(a^2 + b)}{a^2}$$

Next you need to find the denormalization factors:

$$u = 2\pi F_c$$
$$= 6.28 F_c$$

$$\text{ISF} = \frac{F_c}{20\pi}$$

$$= \frac{F_c}{62.8}$$

The capacitance value is denormalized in the usual way

$$C = \frac{C_n}{(u)(\text{ISF})}$$

The normalized capacitance value (C_n) is standardized as 1 F, so

$$C = \frac{1}{(u)(\text{ISF})}$$

Finally, the four resistor values can be denormalized by multiplying the appropriate normalized resistance value by the impedance scaling factor (ISF):

$$R_1 = R_{1n} \times \text{ISF}$$
$$R_2 = R_{2n} \times \text{ISF}$$
$$R_3 = R_{3n} \times \text{ISF}$$
$$R_4 = R_{4n} \times \text{ISF}$$

I'm sure some of this sounds pretty confusing, especially the derivation of intermediate variables a and b, so we work through two examples. You will design an all-pass filter circuit with a gain (K) of 0.25 and a phase shift of $-90°$ at 1 000 Hz (1 kHz).

First, find intermediate variable Y:

$$Y = \tan\left(\frac{w}{-2}\right)$$

$$= \tan\left(\frac{-90}{-2}\right)$$

$$= \tan(45)$$

From Table 7-1 you determine that $Y = 1$.

Next take the gain (K) equation in the form where it is defined in terms of a:

$$a^2 = \left(\frac{1}{K} - 1\right)(1 + a(Y))$$

You know the values of K (0.25) and Y (1), so you can plug these known values into the equation:

$$a^2 = \left(\frac{1}{0.25} - 1\right)(1 + a \times 1)$$
$$= (4 - 1)(1 + a)$$
$$= 3(1 + a)$$
$$= 3a + 3$$
$$0 = a^2 - 3a - 3$$

Do you remember quadratic equations from algebra? Well, that's exactly what you have here. The general form for a quadratic equation is

$$0 = ax^2 + bx + c$$

There are always two possible solutions, which you can identify as s and t:

$$s = \frac{-b + \sqrt{b^2 - 4ac}}{2a}$$

$$t = \frac{-b - \sqrt{b^2 - 4ac}}{2a}$$

In the example, a is the variable to be solved for instead of X. The values of the quadratic variables are

- $a = 1$
- $b = -3$
- $c = -3$

So the value of a is either

$$s = \frac{-(-3) + \sqrt{-3^2 - 4 \times 1 \times -3}}{2 \times 1}$$

192 State variable and all-pass filters

$$= \frac{3 + \sqrt{9 - (-12)}}{2}$$

$$= \frac{3 + \sqrt{9 + 12}}{2}$$

$$= \frac{3 + \sqrt{21}}{2}$$

$$= \frac{3 + 4.58}{2}$$

$$= \frac{7.58}{2}$$

$$= 3.79$$

or

$$t = \frac{-(-3) - \sqrt{-3^2 - 4 \times 1 \times -3}}{2 \times 1}$$

$$= \frac{3 - \sqrt{9 - (-12)}}{2}$$

$$= \frac{3 - \sqrt{9 + 12}}{2}$$

$$= \frac{3 - \sqrt{21}}{2}$$

$$= \frac{3 - 4.58}{2}$$

$$= \frac{-1.58}{2}$$

$$= -0.79$$

Theoretically you can use either s or t as the value for a. Generally speaking, it is a good idea to keep both a and b positive. In fact, the value of b must be positive or you'll design your circuit for negative resistance values. Therefore, choose 3.79 as the value for a. This means the value of b must be

$$b = 1 + a(Y)$$
$$= 1 + 3.79(1)$$
$$= 1 + 3.79$$
$$= 4.79$$

Now you can find the normalized values for the four resistors in the circuit:

$$R_{1n} = \frac{a}{2b}$$
$$= \frac{3.79}{2 \times 4.79}$$
$$= \frac{3.79}{9.58}$$
$$= 0.396 \; \Omega$$

$$R_{2n} = \frac{2}{a}$$
$$= \frac{2}{3.79}$$
$$= 0.527\,7 \; \Omega$$

$$R_{3n} = \frac{2(a^2 + b)}{ab}$$
$$= \frac{2(3.79^2 + 4.79)}{3.79 \times 4.79}$$
$$= \frac{2(14.364\,1 + 4.79)}{18.154\,1}$$
$$= \frac{2 \times 19.154\,1}{18.154\,1}$$
$$= \frac{38.308\,2}{18.154\,1}$$
$$= 2.110\,2 \; \Omega$$

$$R_{4n} = \frac{2(a^2 + b)}{a^2}$$
$$= \frac{2(3.79^2 + 4.79)}{3.79^2}$$
$$= \frac{2(14.364\,1 + 4.79)}{14.364\,1}$$

$$= \frac{2 \times 19.154\,1}{14.364\,1}$$

$$= \frac{38.308\,2}{14.364\,1}$$

$$= 2.667\ \Omega$$

The next step in the design procedure is to find the normalization factors:

$$u = 6.28 F_c$$
$$= 6.28 \times 1\,000$$
$$= 6\,280$$

$$\text{ISF} = \frac{F_c}{62.8}$$
$$= \frac{1\,000}{62.8}$$
$$= 15.923\,6$$

You can round off the impedance scaling factor (ISF) to 16.
Next denormalize the capacitance value:

$$C = \frac{1}{(u)(\text{ISF})}$$
$$= \frac{1}{6\,280 \times 16}$$
$$= \frac{1}{100\,480}$$
$$= 0.000\,009\,9\ F$$
$$= 9.9\ \mu F$$

The next step is to denormalize the four resistor values using the impedance scaling factor (ISF):

$$R_1 = R_{1n} \times \text{ISF}$$
$$= 0.396 \times 16$$
$$= 6.336\ \Omega$$

$$R_2 = R_{2n} \times \text{ISF}$$
$$= 0.527\,7 \times 16$$
$$= 8.443\,2\ \Omega$$

$$R_3 = R_{3n} \times \text{ISF}$$
$$= 2.110\,2 \times 16$$
$$= 33.763\,2\ \Omega$$

$$R_4 = R_{4n} \times \text{ISF}$$
$$= 2.667 \times 16$$
$$= 42.672\ \Omega$$

These resistance values are all far too low to be practical, so multiply all of the resistor values in the circuit by 1 000 and divide the capacitance (C) by the same 1 000:

$$R_1 = 6.336 \times 1\,000$$
$$= 6\,336\ \Omega$$

$$R_2 = 8.443\,2 \times 1\,000$$
$$= 8\,443.2\ \Omega$$

$$R_3 = 33.763\,2 \times 1\,000$$
$$= 33\,763.2\ \Omega$$

$$R_4 = 42.672 \times 1\,000$$
$$= 42\,672\ \Omega$$

$$C = \frac{9.9}{1\,000}$$
$$= 0.009\,9\ \mu F$$

Rounding off to the nearest standard component values, the parts list for the first example all-pass filter circuit looks like

- $C_1, C_2 = 0.01\ \mu F$
- $R_1 = 6.2\ k\Omega\ (6\,200\ \Omega)$
- $R_2 = 10\ k\Omega\ (10\,000\ \Omega)$
- $R_3 = 33\ k\Omega\ (33\,000\ \Omega)$
- $R_4 = 47\ k\Omega\ (47\,000\ \Omega)$

As a second example, design an all-pass filter circuit with a phase shift of $-50°$ at 2 500 Hz (2.5 kHz) and a gain of 0.5. First, find intermediate variable Y:

$$Y = \tan\left(\frac{W}{-2}\right)$$
$$= \tan\left(\frac{-50}{-2}\right)$$
$$= \tan(25)$$

From Table 7-1, you can determine that $Y = 0.445\,2$.

Next take the gain (K) equation in the form where it is defined in terms of a:

$$a^2 = \left(\frac{1}{K} - 1\right)(1 + a(Y))$$

You know the values of K (0.5) and Y (0.445 2), so you can plug these known values into the equation:

$$a^2 = \left(\frac{1}{0.5} - 1\right)(1 + a \times 0.445\,2)$$

$$= (2 - 1)(1 + 0.445\,2a)$$

$$= 1(1 + 0.445\,2a)$$

$$= 0.445\,2a + 1$$

$$0 = a^2 - 0.445\,2a - 1$$

The variables in the quadratic equation are

- $a = 1$
- $b = -0.445\,2$
- $c = -1$

So the value of a is either

$$s = \frac{-(-0.445\,2) + \sqrt{-0.445\,2^2 - 4 \times 1 \times -1}}{(2 \times 1)}$$

$$= \frac{0.445\,2 + \sqrt{0.198\,2 - (-4)}}{2}$$

$$= \frac{0.445\,2 + \sqrt{0.198\,2 + 4}}{2}$$

$$= \frac{0.445\,2 + \sqrt{4.198\,2}}{2}$$

$$= \frac{0.445\,2 + 2.048\,9}{2}$$

$$= \frac{2.494\,1}{2}$$

$$= 1.25$$

or

$$t = \frac{-(-0.4452) - \sqrt{-0.4452^2 - 4 \times 1 \times -1}}{(2 \times 1)}$$

$$= \frac{0.4452 - \sqrt{0.1982 - (-4)}}{2}$$

$$= \frac{0.4452 - \sqrt{0.1982 + 4}}{2}$$

$$= \frac{0.4452 - \sqrt{4.1982}}{2}$$

$$= \frac{0.4452 - 2.0489}{2}$$

$$= \frac{-1.6037}{2}$$

$$= -0.80$$

According to quadratic theory, you can nominally use either s or t as the value for a. Generally speaking, it is a good idea to keep both a and b positive. In fact, the value of b must be positive or you'll design your circuit for negative resistance values. Therefore, choose 1.25 as the value for a. This means the value of b must be:

$$\begin{aligned}b &= 1 + a(Y) \\ &= 1 + 1.25(0.4452) \\ &= 1 + 0.5565 \\ &= 1.5565 \\ &= 1.56\end{aligned}$$

You can get away with rounding off the value of b a little. This rounding off will not change the final results appreciably.

Now you can find the normalized values for the four resistors in the circuit:

$$R_{1n} = \frac{a}{2b}$$

$$= \frac{1.25}{2 \times 1.56}$$

$$= \frac{1.25}{3.12}$$
$$= 0.400\,6 \; \Omega$$

$$R_{2n} = \frac{2}{a}$$
$$= \frac{2}{1.25}$$
$$= 1.6 \; \Omega$$

$$R_{3n} = \frac{2(a^2 + b)}{ab}$$
$$= \frac{2(1.25^2 + 1.56)}{1.25 \times 1.56}$$
$$= \frac{2(1.562\,5 + 1.56)}{1.95}$$
$$= \frac{2 \times 3.122\,5}{1.95}$$
$$= \frac{6.245}{1.95}$$
$$= 3.202\,6 \; \Omega$$

$$R_{4n} = \frac{2(a^2 + b)}{a^2}$$
$$= \frac{2(1.25^2 + 1.56)}{1.25^2}$$
$$= \frac{2(1.562\,5 + 1.56)}{1.562\,5}$$
$$= \frac{2 \times 3.122\,5}{1.562\,5}$$
$$= \frac{6.245}{1.562\,5}$$
$$= 3.996\,8 \; \Omega$$

The next step in the design procedure is to find the normalization factors:

$$u = 6.28 F_c$$
$$= 6.28 \times 2\,500$$
$$= 15\,700$$

The all-pass filter

$$\text{ISF} = \frac{F_c}{62.8}$$

$$= \frac{2\,500}{62.8}$$

$$= 39.808\,9$$

You can round off the value of the impedance scaling factor (ISF) to 40.

Next denormalize the capacitance value:

$$C = \frac{1}{(u)(\text{ISF})}$$

$$= \frac{1}{15\,700 \times 40}$$

$$= \frac{1}{628\,000}$$

$$= 0.000\,001\,5 \text{ F}$$

$$= 1.5 \ \mu\text{F}$$

Then denormalize the resistor values:

$$R_1 = R_{1n} \times \text{ISF}$$
$$= 0.400\,6 \times 40$$
$$= 16.02 \ \Omega$$

$$R_2 = R_{2n} \times \text{ISF}$$
$$= 1.6 \times 40$$
$$= 64 \ \Omega$$

$$R_3 = R_{3n} \times \text{ISF}$$
$$= 3.202\,6 \times 40$$
$$= 128.104 \ \Omega$$

$$R_4 = R_{4n} \times \text{ISF}$$
$$= 3.996\,8 \times 40$$
$$= 159.872 \ \Omega$$

Once again, these resistance values are all a bit low, so rescale the circuit. Multiply all of the resistor values in the circuit by a factor or 500 and divide the capacitance value ($C = C_1 = C_2$) by the same factor of 500:

$$R_1 = 16.02 \times 500$$
$$= 80\,10 \ \Omega$$

$$R_2 = 64 \times 500$$
$$= 32\,000\ \Omega$$

$$R_3 = 128.104 \times 500$$
$$= 64\,052\ \Omega$$

$$R_4 = 159.872 \times 500$$
$$= 79\,936\ \Omega$$

$$C = \frac{1.5}{500}$$
$$= 0.003\ \mu F$$

Rounding off these values to the nearest standard component values gives our completed parts list for the circuit:

- C_1, $C_2 = 0.003\,3\ \mu F$
- $R_1 = 8.2\ k\Omega\ (8\,200\ \Omega)$
- $R_2 = 33\ k\Omega\ (33\,000\ \Omega)$
- $R_3 = 62\ k\Omega\ (62\,000\ \Omega)$
- $R_4 = 82\ k\Omega\ (82\,000\ \Omega)$

Notice that the equations for the design of this all-pass filter circuit only give meaningful results if the gain is quite low. In fact, gains less than unity (attenuation) are generally required for this type of circuit. If you need a higher signal gain, you could follow the all-pass filter with an external amplifier stage.

❖ 8
Voltage-controlled filters

SO FAR ALL OF THE FILTER CIRCUITS WE HAVE BEEN WORKING WITH have been fixed filters. That is, F_c is fixed by the specific design of the filter circuit. To change the cutoff frequency (or center frequency), one or more component values must be changed in the circuit.

In some applications, you might want or need to change F_c without physically redesigning the circuit. The F_c value might need to be dynamically changed while the filter circuit is in operation. A simple passive filter circuit, like the one shown in Fig. 8-1 can be designed for a manually variable cutoff frequency with the addition of a simple potentiometer, as illustrated in Fig. 8-2. The fixed resistor sets the minimum resistance seen by the filter circuit. The effective frequency determining resistance (R) in this circuit is the series combination of the fixed resistor (R1) and the resistance between one end of the potentiometer (R2) and its slider.

To see how this works, suppose this passive low-pass filter circuit is made up of the following component values:

- R_1 = 10-kΩ fixed resistor (10 000 Ω)
- R_2 = 25-kΩ potentiometer (25 000 Ω)
- C_1 = 0.1 μF (0.000 000 1 F)

The value of R in this circuit is always equal to the series combination of R_1 and R_2:

$$R = R_1 + R_2$$

where

R_2 = variable resistance of the potentiometer.

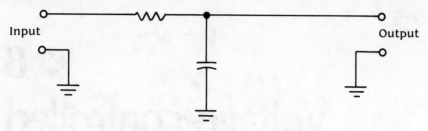

Fig. 8-1 *A simple passive filter network.*

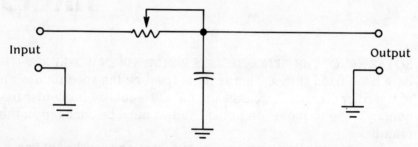

Fig. 8-2 *The circuit of Fig. 8-1 can easily be adapted for a manually controlled cutoff frequency by replacing the fixed resistor with a potentiometer.*

The potentiometer's resistance can be adjusted from 0 Ω (actually a little over 0 Ω, but you can ignore this here) to 25 000 Ω. This means the minimum value of R is

$$R = 10\,000 + 0$$
$$= 10\,000 \text{ Ω}$$

and the maximum value of R is

$$R = 10\,000 + 25\,000$$
$$= 35\,000 \text{ Ω}$$

In other words, R can take on any value between 10 and 35 kΩ. A few possible intermediate values of R are 17 054 Ω, 22 500 Ω, and 29 730 Ω.

The cutoff frequency of this simple passive low-pass filter circuit is determined by the values of R and C as described in chapter 1. The formula for the cutoff frequency (F_c) in this circuit is

$$F_c = \frac{159\,000}{RC}$$

where

F_c = center frequency, in hertz;
R = resistance, in ohms; and
C = capacitance, in microfarads.

In this particular circuit, the value of C is a constant (0.1 μF) but R is variable. Therefore, the actual cutoff frequency (F_c) of the filter can be adjusted by changing the value of R. Because C is a constant, you can rewrite the F_c equation as

$$F_c = \frac{159\,000}{RC}$$

$$= \frac{159\,000}{R \times 0.1}$$

$$= \frac{1\,590\,000}{R}$$

At the minimum setting of potentiometer R2, R has a value of 10 000 Ω, so

$$F_c = \frac{1\,590\,000}{10\,000}$$

$$= 159 \text{ Hz}$$

At the other extreme, when potentiometer R2 is set to its maximum resistance position, R has a value of 35 000 Ω, so the cutoff frequency of the filter becomes

$$F_c = \frac{1\,590\,000}{35\,000}$$

$$= 45.43 \text{ Hz}$$

The cutoff frequency (F_c) of this filter circuit can be manually adjusted from about 45 Hz (when R is at its maximum resistance value) to 159 Hz (when R is at its minimum resistance value). By properly adjusting the setting of potentiometer R2, any frequency between these two extremes can be selected. For example, when R equals 17 454, the cutoff frequency of the filter is

$$F_c = \frac{1\,590\,000}{17\,454}$$

$$= 91.1 \text{ Hz}$$

Manual control over F_c can be easily provided in this way whenever a single resistance element independently controls the cutoff frequency or center frequency of the filter. Unfortunately most of the active filter circuits you have been working with through the last several chapters involve several interrelated component values. You cannot simply change one resistance to alter F_c. Two or more component values must be simultaneously adjusted in perfect unison. Multiple potentiometers mounted on a single control shaft are available, but they are awkward and are generally available only in a very few limited values. If more than two resistances must be simultaneously controlled, a multiple potentiometer rapidly becomes unwieldy.

To make matters worse, some applications require automated control over F_c; that is, the change needs to be made electrically, without an operator physically manipulating a control. This is important in applications requiring rapid changes, precise changes, or precisely timed changes.

A good example of this need is in electronic music synthesizers. Filters are used to shape the timbre of the sound being played. But you want the timbre to sound the same from note to note, with only the pitch changing. Table 8-1 illustrates the problem with fixed filters in such a system. The table shows the effects of a fixed low-pass filter with a cutoff frequency (F_c) of 7 250 Hz (7.25 kHz) on several square-wave signals of various frequencies. A square wave consists of the fundamental frequency and all odd harmonics. In this table, the filter is assumed to be ideal with an infinite roll-off slope. Each frequency component is assumed to be totally present or totally absent in the output. The effects of a practical filter (with a finite roll-off slope) will be similar, though not quite as drastic, at least for frequency components near the nominal cutoff frequency (F_c).

As you can see from Table 8-1, changing the signal's pitch (fundamental frequency) changes the number of harmonics in the output signal and therefore the timbre of the sound. If the signal frequency is higher than 7.25 kHz the entire input signal is filtered out, leaving no output signal. This is obviously undesirable and not a practical musical instrument.

If the musician was adept and daring enough, he might try to manipulate a manual control like on the variable filter circuit described at the beginning of this chapter, but that would be awkward at best. The filter's cutoff frequency would have to be reset for each individual note played. This is probably beyond human capability. Even if it was possible, it would surely be more trou-

Table 8-1 Effects of a fixed filter on a variable frequency signal ($F_c = 7\ 250$ Hz).

Fundamental	Third	Fifth	Seventh	Ninth	Eleventh	Thirteenth	Fifteenth
100 Hz	300	500	700	900	1 100	1 300	1 500
100 Hz	300	500	700	900	1 100	1 300	1 500
300 Hz	900	1 500	2 100	2 700	3 300	3 900	4 500
300 Hz	900	1 500	2 100	2 700	3 300	3 900	4 500
500 Hz	1 500	2 500	3 500	4 500	5 500	6 500	7 500
500 Hz	1 500	2 500	3 500	4 500	5 500	6 500	—
600 Hz	1 800	3 000	4 200	5 400	6 600	7 800	9 000
600 Hz	1 800	3 000	4 200	5 400	6 600	—	—
750 Hz	2 250	3 750	5 250	6 750	8 250	9 750	11 250
750 Hz	2 250	3 750	5 250	6 750	—	—	—
875 Hz	2 625	4 375	6 125	7 875	9 625	11 375	13 125
875 Hz	2 625	4 375	6 125	—	—	—	—
1 000 Hz	3 000	5 000	7 000	9 000	11 000	13 000	15 000
1 000 Hz	3 000	5 000	7 000	—	—	—	—
1 500 Hz	4 500	7 500	10 500	13 500	16 500	19 500	22 500
1 500 Hz	4 500	—	—	—	—	—	—
2 000 Hz	6 000	10 000	14 000	18 000	22 000	26 000	30 000
2 000 Hz	6 000	—	—	—	—	—	—
3 000 Hz	9 000	15 000	21 000	27 000	33 000	39 000	45 000
3 000 Hz	—	—	—	—	—	—	—
7 500 Hz	22 500	37 500	52 500	67 500	82 500	97 500	112 500
—	—	—	—	—	—	—	—

ble than it's worth. Of course, in actual electronic music synthesizers, the process is automated using voltage control.

Basics of voltage control

In a voltage control system, a small, easily controlled electrical signal called the control voltage is fed into a circuit to control one or more parameters of its operation. A very simple and rather crude approach to voltage control is to use an optoisolator. An optoisolator looks like an IC, but it contains a light source and a light sensor in a lighttight housing. A very simple optoisolator is illustrated in Fig. 8-3. The light source is usually an LED. In this case the light sensor is a photoresistor, but any light-sensitive device can be used. Optoisolators are available with photodiodes, phototransistors, and LASCRs as their output devices.

Voltage-controlled filters

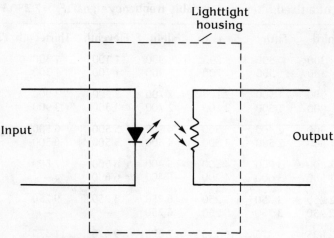

Fig. 8-3 *A simple optoisolator is made up of an LED and a photoresistor in a lighttight housing.*

An external voltage is fed into the pins connected to the internal light source (LED). The higher this applied voltage is, the brighter the LED will glow. Varying the applied voltage will alter the brightness of the light. This varying light level shines on the photosensitive surface of the light sensor (photoresistor). The resistance between the leads of a photoresistor varies in direct proportion to the amount of light shining on its surface at that instant. The optoisolator is enclosed in a lighttight housing so the light sensor won't get "confused" by random light from external (uncontrolled) sources.

Any circuit connected across the output of this optoisolator "sees" a varying resistance. The resistance in the circuit varies in direct proportion to the control voltage fed into the optoisolator's input. The output pins of the optoisolator can be used in place of almost any fixed resistor in virtually any electronic circuit. Whatever operational parameter is affected by that particular resistance will now be under voltage control.

More sophisticated voltage control can be achieved by using an optoisolator with a photodiode or phototransistor output, but in all cases the basic principles are the same. Of course, this won't help much in an active filter circuit where multiple component values combine to determine the cutoff frequency or center frequency (F_c). For active filter circuits, a specialized design is usually required for voltage control rather than crudely converting an existing circuit for voltage control after the fact.

Not surprisingly, a filter circuit that operates under voltage control is known as a voltage-controlled filter (VCF). A voltage-controlled filter changes its cutoff frequency or center frequency (F_c) in response to the applied control voltage, which is usually a low-voltage dc signal (although ac control voltages might occasionally be used in some applications). Some band-pass or band-reject voltage-controlled filters might also have their Q or bandwidth under separate voltage control.

Returning to the electronic music synthesizer from our earlier example, the keyboard puts out a specific dc voltage for each individual note. This dc control voltage is simultaneously fed to a voltage-controlled oscillator (VCO) and a VCF. The VCO changes its frequency (pitch) in response to the control voltage, while the VCF changes its cutoff frequency or center frequency. Because the same control voltage is used for both the VCO and a VCF, they are changed by equal amounts (similar designs for the two circuits are assumed, of course) and the changes are made in step with each other. The output of the filter will have a constant timbre, with the same harmonic content for each note, although the pitch (frequency) changes.

Regeneration and resonance

Some voltage-controlled filters used for electronic music feature an extra control that is usually labelled regeneration or resonance. When this control is set at its minimum position, the filter acts in the ordinary way. As the regeneration or resonance control is advanced towards its maximum position, the F_c (cutoff or center frequency) has a greater emphasis. The frequency response of the filter now resembles an exaggerated Chebyshev response, as illustrated in the graph of Fig. 8-4.

Essentially what is happening is that the output of the VCF is being fed back to its input. If you are using a voltage-controlled filter without a regeneration/resonance control, the effect can be simulated with a mixer, as illustrated in Fig. 8-5.

If the feedback level is sufficient, the voltage-controlled filter will break into oscillations. A strong sine wave with a frequency equal to the filter's F_c will appear at the output, even if no input signal is being fed into the voltage-controlled filter. Because the F_c of a voltage-controlled filter is determined by the control voltage, the VCF becomes a VCO.

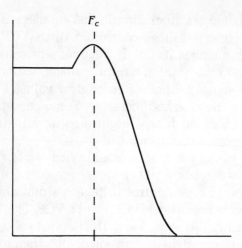

Fig. 8-4 *The frequency response of a filter with regeneration or resonance.*

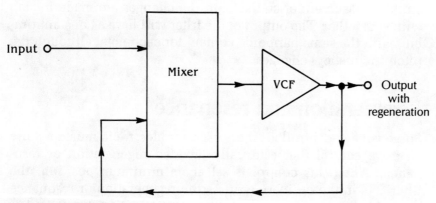

Fig. 8-5 *Regeneration effects in a VCF can be created by feeding some of the output signal back to the filter's input.*

Design of voltage-controlled filters

With a fixed filter, the design procedure focuses on finding the appropriate component values for a specific F_c. In a voltage-controlled filter, the F_c is determined by the externally applied control voltage. The same VCF circuit can be used over a very wide range of F_c values, so individual designs are not needed. You could say, "one design fits all." Therefore, the circuits presented in this chapter won't be as detailed in the design specifics as the fixed filter circuits in the earlier chapters of this book. A great many different types of voltage-controlled filter circuits have

Low-pass VCF

A simple but effective low-pass voltage-controlled filter circuit is shown in Fig. 8-6. Basically, what you have here is a simple active filter circuit with the addition of FET Q1. The signal from the drain of the FET disrupts the signal passing through the filter network changing its operational characteristics. The FET's drain signal directly influences the cutoff frequency (F_c) of the filter. Of course, the FET's drain signal is directly proportional to the externally applied control voltage fed into the FET's gate. In this circuit, the control voltage must be negative.

The main body of the filter is designed for the desired nominal cutoff frequency of the voltage-controlled filter. This will be the maximum F_c for the circuit. The voltage-controlled filter will use this nominal cutoff frequency when the applied control voltage is 0 V (no control voltage). As the control voltage is made more negative (remember, positive control voltages are not permitted in this circuit), the effective cutoff frequency is decreased. The useable range of control voltages in this circuit runs from 0 V to about −5 V.

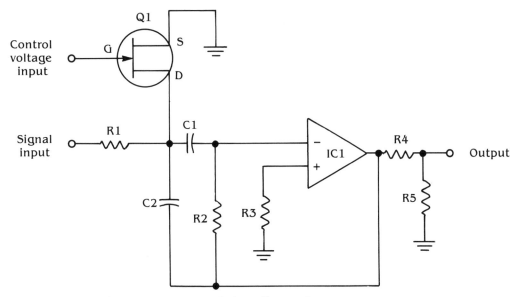

Fig. 8-6 *A very simple but effective low-pass VCF circuit.*

Actually, the core active filter in this circuit is not really a low-pass filter, but an infinite gain multiple feedback band-pass filter circuit. This type of circuit was covered in some detail in chapter 5. In this application, the band-pass filter is designed for a very wide bandwidth with an upper cutoff frequency (F_h) that is above the maximum useful frequency, even when the voltage controlled F_c (actually the lower cutoff frequency (F_1)) is at its minimum. This VCF circuit was designed for use in an electronic music synthesizer, so the only frequencies of interest are those in the audible region (20 Hz to 20 kHz). If the nominal bandwidth extends to 35 to 40 kHz, all audible high frequencies will be preserved in the output, regardless of the control voltage applied to the filter.

Try designing a voltage-controlled filter with a nominal (0 control voltage) cutoff frequency of 5 000 Hz (5 kHz). Remember, this is actually the lower cutoff frequency (F_1) of a band-pass filter. Set the bandwidth at 16 000 Hz (16 kHz) so the entire audible range will be covered regardless of the reduction in F_c due to the applied control voltage.

The actual center frequency, which you need to know for your design, is equal to

$$F_c = F_1 + \left(\frac{BW}{2}\right)$$

$$= 5\,000 + \left(\frac{16\,000}{2}\right)$$

$$= 5\,000 + 8\,000$$

$$= 13\,000 \text{ Hz}$$

$$= 13 \text{ kHz}$$

The Q of this filter is

$$Q = \frac{F_c}{BW}$$

$$= \frac{13\,000}{16\,000}$$

$$= 0.812\,5$$

The gain (K) in this type of circuit is dependent on Q, so

$$K = 2Q^2$$

$$= 2 \times 0.812\,5^2$$

$$= 2 \times 0.660\ 2$$
$$= 1.320\ 4$$

You can round the gain off to 1.3.

Capacitors C1 and C2 are given the same value (C), which for the purposes of design are normalized to 1 F. Resistor R3, which is included in the circuit for improved stability, has the same value as resistor R2. Resistors R4 and R5 form a simple voltage divider network to set the level and impedance of the output signal to suit later stages in the system. Just standardize their values at $R_4 = 100\ \Omega$ and $R_5 = 560\ \Omega$.

This leaves two resistor values (R_1 and R_2) to solve for in the design. Therefore, you must find two conductance values—G_2 and G_1. It is easiest to solve for G_2 first:

$$G_2 = \frac{C}{2Q}$$
$$= \frac{1}{2Q}$$
$$= \frac{1}{2 \times 0.812\ 5}$$
$$= \frac{1}{1.625}$$
$$= 0.615\ 4\ \text{mho}$$

Next, find conductance value G_1:

$$G_1 = 2G_2 K$$
$$= 2 \times 0.615\ 4 \times 1.3$$
$$= 1.6\ \text{mhos}$$

The next step in the design is to find the appropriate normalization factors:

$$u = 2\pi F_c$$
$$= 6.28 F_c$$
$$= 6.28 \times 13\ 000$$
$$= 81\ 640$$

$$\text{ISF} = \frac{F_c}{20\pi}$$
$$= \frac{F_c}{62.8}$$

$$= \frac{13\,000}{62.8}$$

$$= 207$$

Now you can denormalize the capacitor value (C):

$$C = \frac{C_n}{(u)(\text{ISF})}$$

$$= \frac{1}{81\,640 \times 207}$$

$$= \frac{1}{16\,899\,480}$$

$$= 0.000\,000\,06 \text{ F}$$

$$= 0.06 \ \mu\text{F}$$

To denormalize the resistor values, multiply the impedance scaling factory (ISF) by the reciprocal of the appropriate conductance value:

$$R_1 = \frac{\text{ISF} \times 1}{G_1}$$

$$= \frac{\text{ISF}}{G_1}$$

$$= \frac{207}{1.6}$$

$$= 129 \ \Omega$$

$$R_2 = \frac{\text{ISF} \times 1}{G_2}$$

$$= \frac{\text{ISF}}{G_2}$$

$$= \frac{207}{0.615\,4}$$

$$= 336 \ \Omega$$

These resistor values are a bit too low to be practical, so you can multiply them by 1 000 and divide the capacitance value by the same 1 000:

$$R_1 = 129 \times 1\,000$$

$$= 129\,000 \ \Omega$$

$$R_2 = 336 \times 1\,000$$
$$= 336\,000\ \Omega$$
$$C = \frac{0.06}{1\,000}$$
$$= 0.000\,06\ \mu F$$
$$= 60\ pF$$

Rounding off to the nearest standard component values gives us a parts list that looks like this:

- C_1, C_2 = 56 pF
- R_1 = 120 kΩ (120 000 Ω)
- R_2, R_3 = 330 kΩ (330 000 Ω)
- R_4 – 100 Ω
- R_5 = 560 Ω

For improved accuracy, you could place a 10-kΩ (10 000-Ω) resistor in series with R1 for a total resistance of 130 000 Ω, which is closer to the calculated value.

For any serious audio application, a high-grade, low-noise op amp IC should be used. A good choice for the FET is the MPF-102. The exact requirements for the FET are not terribly critical in this circuit (although they will affect the exact control voltage to F_c ratio) and many substitutions are possible.

Band-pass voltage-controlled filter

A somewhat more sophisticated voltage-controlled filter circuit is shown in Fig. 8-7. This is a band-pass VCF. Actually, the core filter section is a state variable filter circuit, so low-pass and high-pass outputs could be tapped off if desired.

Five op amp stages are required in this circuit. Three of them (IC1, IC3, and IC5) are ordinary op amps. The other two (IC2 and IC4), however, are specialized devices known as operational transconductance amplifiers (OTAs).

The symbol for an OTA is illustrated in Fig. 8-8. Notice that this is basically the same as the symbol used to represent an ordinary op amp, except for a minor addition near the output. In somewhat simplified terms, an OTA is an op amp with an extra current input (usually labelled I_{bias}) that controls the gain. Using Ohm's law, we can convert a control voltage into a control current simply by passing it through a resistor, as shown in Fig. 8-9.

214 Voltage-controlled filters

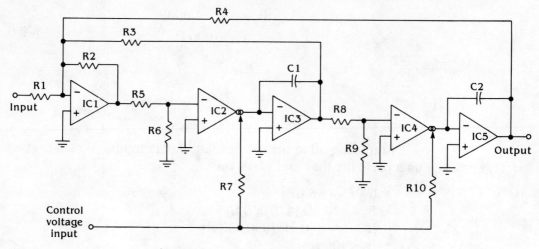

Fig. 8-7 *A more sophisticated VCF circuit.*

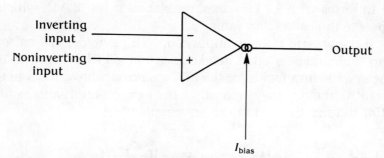

Fig. 8-8 *The operational transconductance amplifier (OTA) can be useful in VCF circuits.*

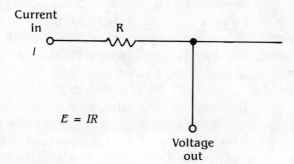

Fig. 8-9 *A control voltage can be converted into a control current by passing it through a resistor.*

By changing the control voltage, the gain of the amplifier is proportionately changed. In other words, an OTA is a simple voltage-controlled amplifier (VCA). Actually, it is a current-controlled amplifier, but you don't have to be particularly concerned over this minor difference.

A typical and popular OTA IC is the CA3080 (see Fig. 8-10). The basic output equation for the CA3080 is:

$$I_o = 19.2 I_c V_i$$

where

I_o = output current,
I_c = control current (fed into the I_{bias} input),
V_i = input voltage, and
19.2 = constant determined by the internal circuitry of the OTA chip.

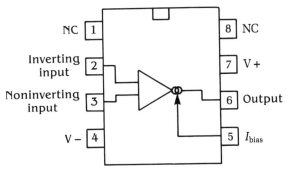

Fig. 8-10 The CA3080 is a popular OTA chip.

Figure 8-11 shows one of the voltage-controlled sections of our complete VCF circuit (from Fig. 8-7). For the time being, concentrate solely on this section of the circuit. The input signal is fed through a simple resistive voltage divider network made up of resistors R5 and R6. Good standardized values for these resistors are

- R_5 = 100 kΩ (100 000 Ω)
- R_6 = 220 Ω

The input signal is reduced by this voltage divider network, so the input voltage (V_i) seen by the OTA is equal to

$$V_i = \frac{R_6 V_x}{R_5}$$

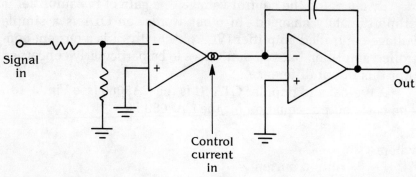

Fig. 8-11 *The voltage-controlled section of the complete VCF circuit of Fig. 8-7.*

$$= \frac{220 V_x}{10\,000}$$

$$= 0.002\,2 V_x$$

where

V_x = original input voltage before it is fed through the voltage divider network.

In operation, the OTA acts like a current-controlled resistance (with I_c as the control current). The equivalent resistance works out to a value of

$$R_{eq} = \frac{V_x}{I_o}$$

$$= \left(\frac{V_i}{0.002\,2} \times \frac{1}{19.2 I_c V_i}\right)$$

$$= \frac{V_i}{(0.002\,2 \times 19.2 I_c V_i)}$$

$$= \frac{V_i}{0.042\,24 I_c V_i}$$

$$= \frac{1}{0.042\,24 I_c}$$

$$= \frac{23.7}{I_c}$$

The control current (I_c) is developed by feeding a control voltage through resistor R7. According to Ohm's law it is equal to

$$I_c = \frac{E_c}{R_7}$$

A good standard value for resistor R7 is 33 kΩ (33 000 Ω), so

$$I_c = \frac{E_c}{33\ 000}$$

Combining these equations, you can state the equivalent resistance in terms of a control voltage instead of a control current:

$$R_{eq} = \frac{23.7}{I_c}$$

$$I_c = \frac{E_c}{33\ 000}$$

$$R_{eq} = \frac{23.7}{\frac{E_c}{33\ 000}}$$

$$= \frac{23.7 \times 33\ 000}{E_c}$$

$$= \frac{782\ 100}{E_c}$$

The center frequency of the filter is equal to

$$F_c = \frac{1}{2\pi C R_{eq}}$$

$$= \frac{1}{6.28 C R_{eq}}$$

$$= \frac{1}{6.28 C \left(\frac{782\ 100}{E_c}\right)}$$

$$= \frac{E_c}{6.28 C \times 782\ 100}$$

$$= \frac{E_c}{4\ 911\ 588 C}$$

For audio work, a good value for C is 0.001 μF (0.000 000 001 F), so

$$F_c = \frac{E_c}{4\,911\,588 \times 0.000\,000\,001}$$

$$= \frac{E_c}{0.004\,912}$$

where

E_c = control voltage.

The second OTA section (IC4, IC5, R8, R9, R10, and C2) is identical to the first.

Looking at the entire circuit (Fig. 8-7), resistors R1, R2, and R4 all have the same value as resistor R5, which is already set to a standard value of 100 kΩ (100 000 Ω). This leaves just one unsolved component in the circuit—resistor R3. The value of this resistor determines the Q of the band-pass filter. This resistor will typically be relatively small, because its value is equal to two times the desired Q:

$$R_3 = 2\,Q$$

In electronic music work, a band-pass filter will usually be set for a fairly high Q. Suppose you want a Q of 50. In this case, the value of resistor R3 is

$$R_3 = 2 \times 50$$
$$= 100\,\Omega$$

A complete parts list for this band-pass voltage-controlled filter circuit is given in Table 8-2.

Table 8-2
Parts list for the VCF circuit of Fig. 8-7.

IC1, IC3, IC5	Low-noise standard op amp
IC2, IC4	OTA (CA3080 or similar)
C1, C2	0.001-μF capacitor
R1, R2, R4, R5, R8	100-kΩ, 0.25-W, 5% resistor
R3	100-Ω, 0.25-W, 5% resistor
R6, R9	220-Ω, 0.25-W, 5% resistor
R7, R10	33-kΩ, 0.25-W, 5% resistor

❖ 9
Equalizers

MANY MODERN SOUND SYSTEMS INCORPORATE AN EQUALIZER OF some sort. An equalizer is nothing but a bank of filters used to "shape" the sound. Generally speaking, an equalizer consists of two or more filters (usually of the band-pass type) in parallel. Each of these paralleled filters controls some specific portion of the audio spectrum. By breaking up the audible spectrum into several slightly overlapping ranges, you can achieve considerable control over the tonal quality of the reproduced sound.

There are two basic types of equalizers. One type is the band equalizer (usually in the form of the graphic equalizer). The second type of equalizer is the parametric equalizer. I briefly consider both types of equalizer in this chapter. I do not deal with specific circuit designs, but I do discuss the basic concepts involved in their specific application. The actual filter circuits used in equalizers are, for the most part, variations on the basic filter circuits described in earlier chapters (particularly chapter 5).

The band equalizer

A band equalizer is sometimes called a fixed filter bank, and that's really all it is. It is a set of band-pass filters wired in parallel. Each individual filter stage has a fixed center frequency and bandwidth selected so that the bands overlap, as illustrated in the combined frequency response graph of Fig. 9-1. Because of the slight overlap in the filter's passbands, there are no gaps.

The gain or attenuation of each filter's passband can be independently controlled to emphasize or deemphasize specific frequency ranges. In equalizer lingo, attenuation (negative gain) of

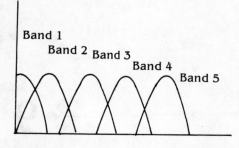

Fig. 9-1 A band equalizer covers the entire audio spectrum in overlapping bands.

a band is called cut and amplification (positive gain) of a band is called boost.

The bass and treble controls on many inexpensive stereos form a supersimple band equalizer. The bass control adjusts the cut or boost of low-end frequencies, while the treble control takes care of the cut or boost of high-end frequencies. Some audio equipment also features a midrange control to handle intermediate frequencies between the bass and treble extremes. The trouble with such bass and treble (and midrange) controls is that their covered passbands are so wide the user can exert only crude control over the tonal quality of the reproduced sound.

A band equalizer will usually have at least five bands, and more are common. Most modern band equalizers divide the audio spectrum into seven to twelve independently controllable bands. Most modern band equalizers are graphic equalizers. A graphic equalizer uses slidepots for the cut/boost controls instead of the more common and familiar rotary potentiometers. As shown in Fig. 9-2, it is very easy to see where a slidepot's slider is positioned. The arrangement of all the equalizer's sliders creates a sort of frequency response graph of the equalizer's effect on the sound passing through it, as illustrated in Fig. 9-3. This is why this type of device is called a "graphic" equalizer. There is nothing special about a graphic equalizer, other than the physical arrangement of its controls.

Graphic equalizers are often used in recording studios, though parametric equalizers (discussed in the next section of this chapter) are generally more useful in such applications. In the last decade or so, graphic equalizers have become increasingly popular accessories in home stereo systems. The equalizer is usually inserted into the system's tape monitor loop, as illustrated in Fig. 9-4. The graphic equalizer will have its own tape monitor inputs/outputs and switch, so an external tape deck can still be used with the system.

The band equalizer

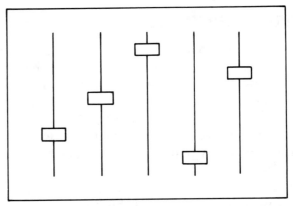

Fig. 9-2 *If a band equalizer uses slidepots, it is a graphic equalizer.*

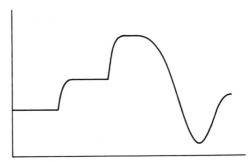

Fig. 9-3 *The positions of the slidepots on a graphic equalizer "draw" a rough graph of the frequency response.*

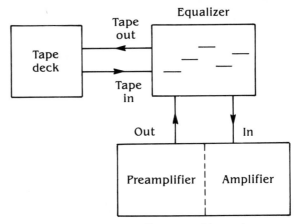

Fig. 9-4 *A graphic equalizer is usually inserted into a home stereo system's tape monitor loop.*

Parametric equalizers

In many serious audio applications, such as in professional recording studios, there isn't any real need for the filter bank to cover the entire audio spectrum, especially in the relatively broad bands of most band equalizers. Instead, in such applications there will often be a need to cut or boost very specific (usually rather narrow) bands of frequencies without affecting adjacent frequencies. A parametric equalizer is used for such a purpose.

A parametric equalizer, as its name suggests, permits user control over various parameters of the individual filter stages in the equalizer. In addition to the cut/boost control, each filter stage will typically have a second control to select the desired center frequency, and a third control to select the desired bandwidth. Only the frequencies that need to be filtered are filtered, unlike a band equalizer that filters everything.

Obviously, a parametric equalizer permits a great deal of control over how it affects the sound. The tradeoff, of course, is its increased complexity. A typical parametric equalizer has at least three controls (boost/cut, center frequency, and bandwidth) for every one (boost/cut only) of a graphic equalizer. Most parametric equalizers have fewer filter stages than graphic equalizers because every frequency in the audio spectrum doesn't have to be covered.

While a few parametric equalizers do use slidepots, they are not positioned like a graphic equalizer. Most parametric equalizers use standard rotary controls that are calibrated by markings on the control panel.

❖ 10
Digital filtering

BY DEFINITION, FILTERING IS AN ANALOG OR LINEAR FUNCTION. But these days digital circuitry is becoming increasingly dominant. Digital circuits have been designed to take over most, if not quite all, standard analog electronic functions. Can a digital circuit act as a filter? Well obviously the answer to that must be yes or there would be no point in this chapter. There are two basic approaches to digital filtering. This chapter covers both the hardware approach and the software approach.

Digital filter circuits

The technical name for a digital filter circuit is a commuting filter. The basic idea behind a commuting filter is illustrated in Fig. 10-1. This is not a functional circuit, but it is included here to illustrate what happens within a real commuting filter circuit.

The input signal is fed through a resistor and one of several switch-selected capacitors. If the switch were permanently left in one position, as shown in Fig. 10-2, you should immediately recognize this circuit as a passive low-pass filter network, just like those you worked with in chapter 1.

But in a commuting filter the switch is not left in just one position. Instead, the switch is continually being moved from capacitor to capacitor at a very rapid (but regular) rate. The switch operates in a rotary fashion. When it passes its last position (the bottommost capacitor) it immediately loops back around to its first position (the uppermost capacitor) and starts the sequence over again.

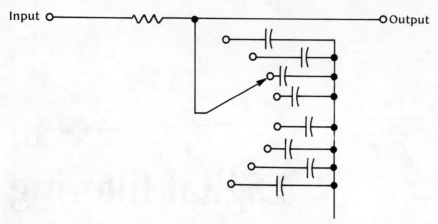

Fig. 10-1 A greatly simplified commuting filter circuit.

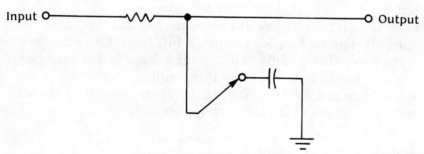

Fig. 10-2 If the switch in Fig. 10-1 is left in one position, we have a simple passive low-pass filter network.

Even though this circuit looks like a low-pass filter, the process of rapidly switching between the capacitors converts the operation of the circuit to something resembling a band-pass filter (although there are some differences which I get to shortly). In a band-pass filter, as you should recall, only a specific range or band of frequencies is passed. Any frequency components outside this passband (either above or below the specified range) are rejected. The basic frequency response graph of a typical band-pass filter is shown in Fig. 10-3. For a commuting filter circuit to work properly, all of the capacitors should have identical values.

The three major variables of interest in the basic commuting filter circuit are R (the resistor value), C (the value of each individual capacitor in the switching circuit), and n (the number of capacitors or switch positions in the sequence). These three variables combine to define the center frequency of the main pass-

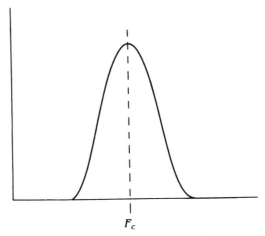

Fig. 10-3 *The frequency response graph for a typical band-pass filter.*

band of the filter according to the equation

$$F_c = \frac{1}{2nRC}$$

The bandwidth is quite narrow for this type of filter.

Try an example. Suppose you have a commuting filter circuit made up of a 10-kΩ (10 000-Ω) resistor and eight 0.01-μF capacitors. In this case, the basic center frequency of the circuit will be equal to

$$F_c = \frac{1}{2 \times 8 \times 10\,000 \times 0.000\,000\,01}$$

$$= \frac{1}{0.001\,6}$$

$$= 625 \text{ Hz}$$

Unlike an ordinary analog band-pass filter, a commuting filter actually has multiple passbands. Harmonics (integer multiples) of the filter's nominal center frequency (F_c as designed from the equation described above) also pass through to the output. For instance, in the example, F_c is equal to 625 Hz, so there will be additional passbands centered around the following frequencies:

- 1 250 Hz (second harmonic)
- 1 875 Hz (third harmonic)
- 2 500 Hz (fourth harmonic)

- 3 125 Hz (fifth harmonic)
- 3 750 Hz (sixth harmonic)
- 4 375 Hz (seventh harmonic)

Each successive harmonic passband has a lower amplitude than its predecessor. In the example, the gain at 2 500 Hz (fourth harmonic) is less than the gain at 1 875 Hz (third harmonic), but greater than the gain at 3 125 Hz (fifth harmonic). Eventually, the amplitude of the high harmonic passbands will be too weak to be of any significance in the output signal. The upper harmonics are therefore effectively filtered out of the signal.

Figure 10-4 shows the frequency response graph for a typical commuting filter. Because the multiple passbands on this graph somewhat resemble the teeth of a comb, the commuting filter is sometimes called a comb filter. If the intended application requires a more traditional bandpass operation (that is, just a single passband without the harmonic passbands), a simple (analog) low-pass filter network at the output of the commuting filter can help get rid of most of the harmonic passbands.

You already know that the passive version of the commuting filter shown in Fig. 10-1 is not a functional circuit. This is because the switch must be continually rotated through its sequence at a steady and rapid rate. Even if an operator tried to turn the switch, he couldn't possibly turn it fast enough, at least not for long. To make a functional commuting filter circuit, you need some way to automatically (electronically) control the switch. This is where the digital circuitry comes in.

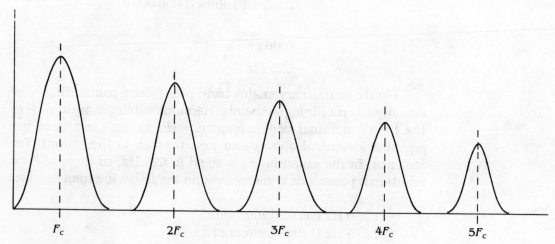

Fig. 10-4 *The frequency response graph for a typical commuting filter.*

A steady square-wave signal, called the clock, is fed through a digital counter. The output of the counter is then decoded by an n decoder of some sort. (Of course, n is the number of capacitors or switch positions in the sequence.) A block diagram of a digital commuting filter circuit is shown in Fig. 10-5.

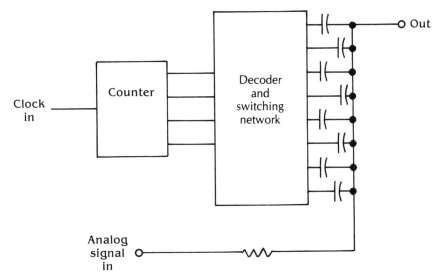

Fig. 10-5 *A block diagram of a digital commuting filter circuit.*

Because the number of capacitors (switch positions) in the sequence affects the nominal center frequency (F_c) of the filter, a digital circuit can be designed to vary the number of sequence steps (n) under digital control. This can be used to automate the center frequency of the filter in an advanced system.

Advantages of software filters

It is also possible to perform filtering operations with digital software. Software is the general term for computer programs—the instructions that tell a computer what to do and when. There are several advantages to filtering software. Any characteristic of the software filter can be changed by rewriting a few lines of programming code. One basic programmable digital circuit can do many different jobs, functioning as many individual dedicated (hardware) circuits. To change the function, you don't have to change any of the physical components in the circuit, you only have to change the programming software.

In addition, virtually any desired filter frequency response pattern can be created. With dedicated circuits, you are limited in your choices. You can build filters with low-pass, high-pass, band-pass, band-reject, and even comb responses. But if you needed an irregular filter response, such as the one illustrated in the graph of Fig. 10-6, you would have a real problem if you had to design a dedicated filter circuit. Such a circuit would be very difficult to design. It might even be impossible. In a software filter, however, there wouldn't be any problem at all. The irregular frequency response pattern of Fig. 10-6 would be significantly more difficult to achieve than a standard low-pass response.

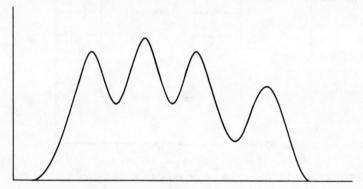

Fig. 10-6 *A software filter can be used to create unusual frequency response patterns.*

A software filter's roll-off slope can be made as steep or as shallow as your particular application requires. In a band-pass filter, for example, different roll-off slopes could be set up for the lower and upper cutoff frequencies, if this were appropriate to your application. (I can't really imagine why you'd want to do this, but the important thing here is that you can do it.)

Digital signals

The primary limitation of software filters is the fact that, like any other digital system, they can only work on digital signals. A digital signal can take on just one of two possible states, which can be identified in various ways:

- Low or high
- Off or on

- No or yes
- 0 or 1

Analog signals, on the other hand, can take on any of an infinite number of possible values within a given range. Between any two analog values A and B, there are always additional values midway between A and B. For example, between 1.0 and 2.0, you can have 1.5. Between 1.5 and 2.0 you can have 1.75. Between 1.5 and 1.75 you can have 1.625. This can go on indefinitely. But digital signals have no intermediate values. A digital signal must be either 0 or 1. It can never be between 0 and 1.

Each digital signal is called a bit. A single bit can't carry much information because it has just two possible states. To carry more information, digital bits are combined into multidigit words. A digital word made up of eight bits is called a byte, and a four-bit word is called a nibble. Using more bits allows greater detail in the carried data.

As an example, if you use two bits, you have four possible combinations:

- 0 0 (0)
- 0 1 (1)
- 1 0 (2)
- 1 1 (3)

If you add a third bit, you can have up to eight possible combinations. Four-bit words (nibbles) offer 16 possible combinations, as outlined in Table 10-1.

As you can see, each digital word is basically just a number. A special system for counting is used for these digital signals.

**Table 10-1
Four-bit nibbles
and their decimal values.**

0000	0	1000	8
0001	1	1001	9
0010	2	1010	10
0011	3	1011	11
0100	4	1100	12
0101	5	1101	13
0110	6	1110	14
0111	7	1111	15

230　Digital filtering

This is called the binary number system. The word binary means there are two available digits in this system: 0 and 1. We are more familiar with the decimal numbering system which has ten available digits: 0, 1, 2, 3, 4, 5, 6, 7, 8, and 9. All you really need to know about the binary numbering system is that it's the way a computer (or other digital circuit) counts, and it is made up of nothing but a bunch of zeroes and ones.

A/D and D/A conversion

To use a software filter with an analog signal, you have to convert the analog signal into a string of binary numbers that can be recognized and manipulated by the digital circuitry and its software. This process is called analog to digital (A/D) conversion. A circuit for accomplishing this is called an A/D converter.

There are many different types of A/D converter circuits and I won't go into the specific details here. Basically, they all sample the signal several times per cycle and convert the instantaneous voltage at each sample into an appropriate numerical (digital) value. Because the digital circuitry doesn't permit intermediate values, each sample value must be rounded off to the nearest whole unit. The process is illustrated in Fig. 10-7. The result of the A/D conversion is a repeating series of numbers corresponding to the original analog waveform.

To return to the analog realm for reproduction of the sound

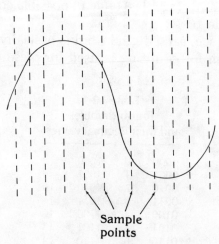

Fig. 10-7 In A/D conversion, each analog sample value is converted into a whole number digital value.

(or other analog signal), you use a process called digital to analog (D/A) conversion. The circuit used for this is known as a D/A converter. The digital numbers are fed through the D/A converter in sequence. As each digital number passes through the circuit, it produces a proportional analog voltage at the output. By stepping through the pattern of digital values at the correct rate, the changing voltages at the output will recreate the original analog waveform, as illustrated in Fig. 10-8.

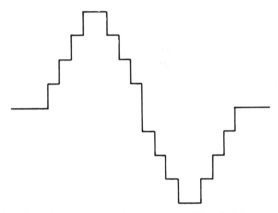

Fig. 10-8 *The digital values can be converted back into an analog waveform.*

Usually a D/A converter will be followed by a simple low-pass filter circuit of some sort (often a passive filter network will be used here). The function of this filter is to smooth out the sharp corners of the steps between adjacent voltage values, as illustrated in Fig. 10-9.

A digital number is a digital number, so you don't necessarily have to start out with an analog signal. A computer can be programmed to create a pattern of numbers that can then be passed through a D/A converter to create a new analog waveform. This is called a digital oscillator.

The specific numerical relationships between adjacent values determine the waveshape. A typical square-wave pattern is outlined in Table 10-2, Table 10-3 gives the pattern for a triangular wave, and Table 10-4 suggests a pattern for an ascending sawtooth wave. To change the signal frequency, certain steps in the pattern can be omitted or repeated to control the length of the sequence. The step rate (how many numbers are treated per second) is constant.

232 Digital filtering

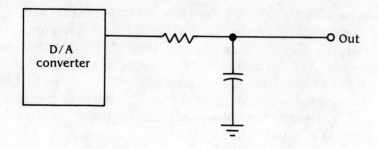

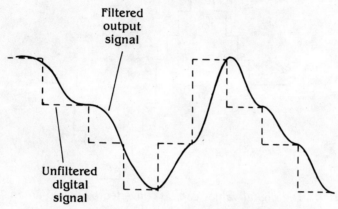

Fig. 10-9 *A simple low-pass filter is used to smooth out the signal from a D/A converter.*

−8
−8
−8
−8
−8
−8
−8
−8
−8
10
10
10
10
10
10
10
10
10
−8

**Table 10-2
Typical software
table for a square wave.**

Table 10-3　Typical software table for a triangular wave.	Table 10-4　Typical software table for a sawtooth wave.
0	0
−2	−8
−4	−7
−6	−6
−8	−5
−6	−4
−4	−3
−2	−2
0	−1
2	0
4	1
6	2
8	3
10	4
8	5
6	6
4	7
2	8
0	0

Software filters

Because the pattern of numerical values making up the waveform in the computer is in digital form, it can be mathematically treated like any other series of numbers. By performing a specific equation on each number in the sequence, the waveshape can be altered in some specific way. This is how software filtering is accomplished. A software filter is nothing but a repeating mathematical operation (or series of operations) that alters the waveform sequence pattern.

For digital filtering calculations, you need to set up a special variable called an accumulator. I identify the accumulator variable as R. The accumulator functions in a manner similar to an analog integrator or low-pass filter. Numbers are repeatedly added to the accumulator for each sample value, so the value of this variable continuously changes.

To simulate a simple passive low-pass RC-type filter, you need to incorporate the effects of capacitive leakage. A constant proportion of the accumulator's value is subtracted from each successive sample to simulate the leakage of an analog, real-world capacitor. The leakage constant is called K.

In an analog filter circuit, the capacitive leakage is directly related to the capacitance value. The cutoff frequency is also dependent on the size of the capacitor. Similarly, in the digital software filter, the cutoff frequency is determined by the value of K.

For each new sample in the waveform sequence (V), you can calculate the accumulator's new value with the formula

$$R = R - (K \times R) + V$$

For the first sample in the sequence, R is assumed to have a previous value of zero.

A new waveform table (or sequence of numbers) is created by applying this equation to each of the old table entries. The result will be the low-pass filtered equivalent of the original signal.

Other similar equations can be used to simulate other filter types. Because almost any mathematical formula can be used, a digital software filter is not limited to the four basic analog filter types (low-pass, high-pass, band-pass, and band-reject). All sorts of unique frequency responses can be set up by employing the appropriate filtering equation.

Index

A
active filters (*see* filters, active)
all-pass filters, 185-200
    formulas, 188-200
    schematic diagram, 186
amplifiers, 56 (*see also* op amps)
analog signals, 229
analog-to-digital (A/D) conversion, 230-232

B
band equalizer, 219-221
band-pass filters, 4
    active, 107-132
    basics, 26-31
    formulas, 29-33, 108-109
    high-order, 132
    infinite gain multiple feedback, 109-120
    passive, 33-41
    passive formulas, 34-41
    passive schematic diagram, 33, 47
    VCVS, 120-126
    VCVS high-Q, 126-131
    voltage-controlled, 213-218
band-reject filters, 5, 41-46
    active, 133-160
    active bridged differentiator schematic diagram, 147
    active subtraction-type, 149-154
    active subtraction-type formulas, 150-154
    active subtraction-type schematic diagram, 149
    active twin-T, 138-142
    active twin-T formulas, 139-142
    active twin-T schematic diagram, 139
    active VCVS, 154-160
    formulas, 45-46, 134-135
    frequency response graph, 134
    passive, 46-49
    passive bridged differentiator, 142-148
    passive bridged differentiator formulas, 143-148
    passive bridged differentiator schematic diagram, 143, 146
    passive formulas, 47-49
    passive schematic diagram, 47, 136
    passive twin-T, 135-138
    passive twin-T formulas, 136-138
    passive twin-T schematic diagram, 136
    state variable, 176-185
binary numbering system, 229-230
bit, 229
boost, 220
Butterworth filters, 93-98
    active low-pass, 62-64
    active low-pass frequency response graph, 63-64
    active low-pass infinite gain multiple feedback, 66-71
    active low-pass schematic diagram, 64
byte, 229

C

capacitive reactance, 8-11
  examples, 11
  formula, 9-11
center frequency, 133
Chebyshev filters, 93-98
  active low-pass, 65
  active low-pass infinite gain multiple feedback, 73
  active low-pass frequency response graph, 65
chips
  741, 57-58
  CA3080, 215
clock, 227
comb filters (see commuting filters)
commuting filters, 223-227
  digital, 227
  formula, 225
  frequency response graph, 226
cut, 220
cutoff frequency, 15, 133
  relationships between high/low, 26-28
cutoff slope, 63, 86
cycle, 2
cycles per second (cps), 2

D

decibel (dB), 7
decimal numbering system, 230
differential amplifier, 60
differentiators, 91
digital counter, 227
digital filters (see commuting filters)
digital oscillator, 231
digital signals, 228-230
digital-to-analog (D/A) conversion, 230-232

E

equalizers, 219-222
equations (see formulas)

F

filters
  active, 5
  all-pass, 185-200
  band-pass, 4, 33-41
  band-pass active, 107-132
  band-pass basics, 26-31
  band-pass high-order, 132
  band-pass infinite gain multiple feedback, 109-120
  band-pass VCVS, 120-126
  band-pass VCVS high-Q, 126-131
  band-pass voltage-controlled, 213-218
  band-reject, 5, 41-46
  band-reject active, 133-160
  band-reject active subtraction-type, 149-154
  band-reject active twin-T, 138-142
  band-reject active VCVS, 154-160
  band-reject passive, 46-49
  band-reject passive bridged differentiator, 142-148
  band-reject passive twin-T, 135-138
  band-reject state variable, 176-185
  basic principles, 1-14
  basic types, 3-5
  Butterworth active low-pass, 62-64
  Butterworth active low-pass infinite gain multiple feedback, 66-71
  Chebyshev active low-pass, 65
  comb (see commuting filters)
  commuting, 223-227
  comparison, 49-53
  digital (see commuting filters)
  high-order, 86-87
  high-pass, 3-4
  high-pass active, 89-106
  high-pass infinite gain multiple feedback, 93-98
  high-pass passive, 21-25
  high-pass VCVS, 99-106
  Key (see VCVS low-pass filters)
  low-pass, 3
  low-pass active, 55-87
  low-pass advantages/disadvantages, 18-19
  low-pass infinite gain multiple feedback, 66-73
  low-pass passive, 14-19
  notch (see band-reject filters)
  passive, 5, 21-53
  phase-shift (see all-pass filters)
  practical vs. ideal, 6-8
  Salen (see VCVS low-pass filters)

Index 237

software, 233-234
software advantages, 227-228
state variable band-reject, 176-185
state variable basics, 161-185
state variable four op amp, 168-176
state variable unity gain, 163-168
symbols, 25
universal (see state variable)
VCVS band-pass, 120-126
VCVS band-pass high-Q, 126-131
VCVS band-reject active, 154-160
VCVS high-pass, 99-106
VCVS low-pass, 73-86
voltage-controlled (VCFs), 201-218
fixed filter bank (see band equalizer)
formulas
  all-pass, 188-200
  band-pass, 29-33
  band-pass active, 108-109
  band-pass infinite gain multiple feedback, 111-120
  band-pass passive, 34-41
  band-pass VCVS, 121-126
  band-pass VCVS high Q, 127-131
  band-pass voltage-controlled, 215-218
  band-reject, 45-46
  band-reject active subtraction-type, 150-154
  band-reject active twin-T, 139-142
  band-reject active VCVS, 155-160
  band-reject passive, 47-49
  band-reject passive bridged differentiator, 143-148
  band-reject passive twin-T, 136-138
  band-reject state variable, 177-185
  Butterworth active low-pass infinite gain multiple feedback, 67-73
  capacitive reactance, 9-11
  commuting filters, 225
  G_2, 67-73
  G_2, calculating directly, 71-73
  high-order, 87
  high-pass infinite gain multiple feedback, 95-98
  high-pass passive, 23-25
  high-pass VCVS, 100-106

  inverting amplifier, 60
  low-pass, passive, 15-16
  low-pass VCVS, 75-78
  low-pass VCVS direct design, 80-86
  low-pass VCVS equal component method, 78-80
  RC time constants, 13-14
  state variable band-reject, 177-185
  state variable four op amp, 169-176
  state variable unity gain, 163-168
  voltage-controlled filters (VCFs), 201-203, 210-213
frequency, 1-2
  fundamental, 49, 90
  types, 49
frequency response graphs, 4-5, 27-28, 43-44
  band-reject, 134
  Butterworth, 63-64
  Chebyshev active low-pass, 65
  commuting, 226
  high-pass passive, 22

G

gain, 60-62
graphic equalizer, 219

H

harmonics, 49-53, 90, 204-205, 225-226
hertz (Hz), 2
high-order filters, 86-87
  formulas, 87
high-pass filters, 3-4
  active, 89-106
  active noise, 92-93
  infinite gain multiple feedback, 93-98
  passive, 21-25
  passive formulas, 23-25
  passive frequency response graph, 22
  passive schematic diagram, 22
  symbol, 25
  VCVS, 99-106

I

infinite gain multiple feedback
  band-pass, 109-120

infinite gain multiple feedback (cont.)
   band-pass formulas, 111-120
   band-pass schematic diagram, 110, 114
   Butterworth active low-pass formulas, 67
   Chebyshev active low-pass filters, 73
   high-pass, 93-98
   high-pass formulas, 95-98
   high-pass schematic diagram, 95
   low-pass schematic diagram, 66
insertion loss, 55
integrators, 91
inverting amplifier
   formula, 60
   schematic diagram, 61
inverting voltage follower, 61

K

Key filters (see VCVS low-pass filters)
kilohertz (kHz), 2

L

linear scale, 8
logarithmic scale, 8
low-pass filters, 3
   active, 55-87
   active Butterworth, 62-64
   active Chebyshev, 65
   advantages/disadvantages, 18-19
   infinite gain multiple feedback, 66-73
   passive, 14-19
   passive formulas, 15-16
   passive schematic diagram, 22
   symbol, 25
   VCVS, 73-86

M

megahertz (MHz), 2

N

noise, active high-pass filters, 92-93
noninverting amplifier, schematic diagram, 62
notch filters (see band-reject filters)
nibble, 229

O

octave, 7
Ohm's law, 217
op amps, 56-62 (see also chips)
   symbol, 59
operational amplifiers (see op amps)
operational transconductance amplifiers (OTAs), 214
optoisolator, 205-207
order, 63
overtone frequency, 49

P

parametric equalizer, 219, 222
passive filters (see filters, passive)
phase-shift filters (see all-pass filters)
pi symbol, 36

Q

Q factor, 31-33, 108-109
quadratic theory, 197
quality factor (see Q factor)

R

RC network, 12-14
   combinations, 17
RC time constants, 12-14
   formula, 13-14
regeneration, 207-208
resonance, 207-208

S

Salen filters (see VCVS low-pass filters)
sawtooth wave, 233
   harmonic contents, 50-53
schematic diagrams
   all-pass filter, 186
   band-pass infinite gain multiple feedback, 110, 114
   band-pass passive, 33, 47
   band-pass VCVS, 121
   band-pass VCVS high-Q, 127
   band-pass voltage-controlled, 214
   band-reject active bridged differentiator, 147

band-reject active subtraction-type, 149
band-reject active twin-T, 139
band-reject passive, 47, 136
band-reject passive bridged differentiator, 143, 146
band-reject passive twin-T, 136
band-reject state variable, 176, 178
band-reject VCVS active, 154
Butterworth active low-pass, 64
high-pass infinite gain multiple feedback, 95
high-pass passive, 22
high-pass VCVS, 99
inverting amplifier, 61
low-pass infinite gain multiple feedback, 66
low-pass passive, 22
low-pass VCFs, 209
low-pass VCVS, 75
noninverting amplifier, 62
state variable band-reject, 176, 178
state variable four op amp, 169
state variable unity gain, 163
sine wave, 2, 49
slope, 7
software filters, 233-234
  advantages, 227-228
square wave, 49, 90, 232
  frequency components, 50
state variable filters
  band-reject, 176-185
  band-reject formulas, 177-185
  band-reject schematic diagram, 176, 178
  basics, 161-185
  four op amp, 168-176
  four op amp formulas, 169-176
  four op amp schematic diagram, 169, 184
  unity gain, 163-168
  unity gain formulas, 163-168
  unity gain schematic diagram, 163
supply rails, 58
symbols
  high-pass filters, 25
  low-pass filters, 25
  op amp, 59
  pi, 36

T
tangent values, 187
triangular wave, 233
Tschebyscheff (*see* Chebyshev)
Tschebyshev (*see* Chebyshev)

U
undertone frequency, 49
unity gain
  VCVS low-pass, 74-78
  state variable, 163-168
universal filters (*see* state variable filters)

V
VCVS filters
  active band-reject, 154-160
  active band-reject formulas, 155-160
  active band-reject schematic diagram, 154
  band-pass, 120-126
  band-pass formulas, 121-126
  band-pass high-Q, 126-131
  band-pass high-Q formulas, 127-131
  band-pass high-Q schematic diagram, 127
  band-pass schematic diagram, 121
  high-pass, 99-106
  high-pass formulas, 100-106
  high-pass schematic diagram, 99
  low-pass, 73-86
  low-pass direct design formula method, 80-86
  low-pass equal component formula method, 78-80
  low-pass formulas, 75-78
  low-pass schematic diagram, 75
  low-pass unity gain, 74-78
voltage control, basics, 205-207
voltage-controlled amplifiers (VCAs), 215
voltage-controlled filters (VCFs), 201-218
  band-pass, 213-218
  band-pass formulas, 215-218
  band-pass schematic diagram, 214

voltage-controlled filters (VCFs) (*cont.*)
    design, 208-209
    formulas, 201-203, 210-213
    low-pass, 209-213
    low-pass schematic diagram, 209
voltage-controlled oscillator (VCO), 207

voltage-controlled voltage source (*see* VCVS filters)

W

waveshapes (*see* specific types of, i.e., sine, square, etc.)